Desert Warfare: German Experiences in World War II

by
Major General Alfred Toppe

U.S. Army Command and General Staff College
Fort Leavenworth, Kansas 66027-6900

Editor's note: This Special Study is an excerpt from *German Experiences in Desert Warfare During World War II*, by Generalmajor Alfred Toppe and 9 others [et al.], 2 vols., 1952. These 2 vols. include 380 pages, 36 sketches, 15 maps, and 85 photographs. The original study was MS. no. P-129 in the Foreign Studies Series of the Historical Division, United States Army, Europe (now found in the National Archives). Mr. H. Heitman edited and translated the origin manuscript, which has been reformatted and edited, in part, by Combat Studies Institute editors.

CONTENTS

Preface vii
Introduction ix
Chapter I. Prior Planning 1
   1. Intelligence Planning 1
      a. Desert Terrain and Climate 1
      b. Scope of the Evaluation 3
      c. Influence of Intelligence on Planning 3
      d. Availability and Evaluation of Terrain Intelligence 3
      e. Use of Historical Data for Planning Purposes 4
   2. Operational Planning 4
      a. General 4
      b. Changes in Troop Organization and Equipment 7
      c. Special Training 10
      d. Acclimatization of the Troops 12
      e. Development of Special Equipment 13
   3. Logistical Planning 13
      a. February—May 1941 14
      b. June—December 1941 15
      c. January—June 1942 15
      d. July 1942—May 1943 15
Chapter II. Operations 17
   4. General Description of the Zone of Operations 17
      a. Mountain Ranges 18
      b. Steep Terraces 20
   5. Order of Battle of Army and Luftwaffe Units 24
   6. Reasons for Changes in Organization and Equipment 28
   7. Descriptions of the More Important Battles 29
      a. 31 March—19 April: The First Counterattack to Reconquer the Cyrenaica 29
      b. May—June 1941: Battle for the Positions on the Border 32

iii

- c. July—Mid-November: The Siege of Tobruk and Preparations for the Attack 34
- d. Mid-November 1941—Mid-January 1942: Repelling The British Autumn Offensive and the Retreat to the Gulf of Sirte 37
- e. Mid-January—End of May 1942: The Counteroffensive to Retake the Cyrenaica and the Preparations for the Attack on Tobruk 41
- f. Late May—July 1942: The Battle of Tobruk and the Pursuits to El Alamein 44
- g. August—Early November 1942: The Battles Around Alamein 47
- h. November 1942—January 1943: The German Retreat to the Border Between Libya and Tunisia 50
- i. November 1942—March 1943: The Occupation of Tunisia and the Battles Fought in Tunisia 52
- j. April—May 1943: The Final Battle in Tunisia 56

Chapter III. Special Factors 59
- 8. Dust 59
  - a. Effect on Troops, Weapons, and Equipment 59
  - b. Effect on Combat Operations 60
  - c. Effect on Tactical Measures 61
  - d. Effect on Aircraft and Their Crews 63
- 9. Terrain 64
  - a. Influence on Tactical Measures 64
  - b. Influence on the Construction of Field Fortifications and the Use of Weapons 68
  - c. The Tactical Importance of the Recognition of Vehicle Tracks by Air Observation 71
  - d. The Use of Vehicle Tracks for Deception of the Enemy 71
  - e. The Use of Wheeled and Track Vehicles 71
  - f. Influence of Desert Terrain on the Development of New Tactical Principles for the Use of Motorized Units 72
  - g. Influence of Rainfall on Mobility in Desert Terrain 74

10. Water 75
    a. General 75
    b. Requirements for Troops and Vehicles,
       Economy Measures, etc. 75
    c. Water and Motor-Fuel Requirements 76
    d. Tactical Importance of the Presence of Water Sources .. 76
    e. Assignment of Engineer Troops for Water-Supply
       Services 79
    f. Well-Drilling Equipment 79
    g. Method of Distribution 80
    h. Pipelines 81
11. Heat 81
    a. General 81
    b. Effect on Unaccustomed Troops 81
    c. Effect on Tank Crews 82
    d. Measures Taken to Avoid the Noonday Heat 82
    e. Special Equipment for Protection Against
       Temperature Variations 82
    f. Types of Shelter 82
    g. Comparison Between the Efficiency of Troops
       in the Tropics and in Temperate Climates 82
    h. Effect on Materiel and Equipment 83
    i. Effect on Visibility 83
    j. Effect on Airplanes in Taking Off and Landing 83

Chapter IV. Miscellaneous 85
12. Cartographic Service 85
    a. General 85
    b. Reliability and Methods of Use 85
13. Camouflage 86
14. Evaluation of the Enemy Situation Through
    Aerial Photographs 87
15. Visibility at Night 87
16. Choice of Camp Sites 87

v

17.	Selection of Battle Sites	88
18.	Time of Day Selected for Combat	88
19.	Influence of the Desert Climate on Daily Service Routine	88
20.	Special Problems of the Technical Services	89
21.	Influence of Light, Shade, and Sandstorms on Combat	89
22.	Influence of Darkness on Radio Communications	90
23.	Wind	90
24.	Special Equipment and Procedures for Aircraft Crews	90
25.	Dry Docks and Port Installations	92
26.	Reinforcement of Sand Surfaces for Landings by Amphibious Craft	92
27.	Changes in Ship Loading and Unloading Procedures	93
28.	Materiel Losses and Replacement Estimates for Desert Warfare	93
29.	Modifications in Supply-Dump Procedures— Especially for POL	93
30.	Diseases and Insects in the Desert	93
31.	Desert Weather Service	95

Chapter V. General Remarks and Experiences 97

32.	Special Equipment for Desert Warfare	97
33.	Research and Development Possibilities for Special Desert Equipment	99
34.	Unusual Supply Problems	99
35.	Nutrition	100
36.	Clothing	101
37.	Comparison with Desert Warfare in Southern Russia	101
38.	Troop Welfare in the Desert	102

PREFACE

In spite of the time limit imposed on him, Major General Alfred Toppe, the topic leader, with the collaboration of the leading German experts on the African campaign, has succeeded in this work in answering the assigned questions. The esprit de corps and the justified pride of the African veterans were a decided factor that helped to make the contributions so good and comprehensive that they could, to a large extent, be fitted into the attached study. This in no way detracts from the services of the topic leader. It was his initiative and organizational ability that resulted in this excellent study, despite the time restriction.

The German experiences in African desert warfare are made unique by the fact that the command and the troops were faced with a mission in no way either planned or prepared, and they entered into it completely without prior prejudices. The experience gained, therefore, is free from outside theories and opinions and was only achieved by their struggling with an entirely new military situation; it thus has the value of originality. The value is diminished, however, by the fact that the experiences are in part negative and could not be developed further in a positive direction due to the lack of time and the limited means at hand.

The particular conditions in Africa under which they were gained will have to be kept in mind in any evaluation. The impossibility of securing a supply line across a body of water dominated by the enemy, the numerical and material inadequacies held by the Germans—and even more their allies—and the increasing lack of Luftwaffe fighting and transport units—these are all negative aspects of the campaign. On the positive side belongs the tempo and performance of field forces under the leadership of Rommel, forces which were without a doubt far above the average in initiative, spontaneity, and soldierly zeal.

Generaloberst Franz Halder

Koenigstein/Taunus
18 June 1952

INTRODUCTION

Two and a half months was the total time allotted for the preparation of this study. Prerequisite was that such German officers be induced to contribute who had had as broad as possible a view in the conduct of overall operations, who possessed practical combat experience, and, furthermore, who had exact knowledge of as many factors as possible that exerted a determining influence on desert warfare. In addition to the contributors listed below, a number of former members of the German Africa Corps also made contributions.

The organization of this study was based on the individual questions assigned; German manuals were not used. The presentation, therefore, can be evaluated on the basis of actual experiences.

A number of questions could not be answered exhaustively. The reason for this lies in the fact that no experience had been gathered in such areas, or else operations took place in areas in which the typical attributes of a real desert were not present. The request (attached to the major question) that accounts by "individuals or groups" be added concerning "Special Equipment and Procedures for Aircraft Crews" could not be fulfilled because no authorities on this subject could be contacted in the short time available. A broad survey of important battles has been included in chapter II, section 7. The official documents contained in Field Marshal Rommel's notes also have been utilized as a valuable source of information.

Those who have contributed information and analysis to this study include:

> Major General (Generalleutnant) Fritz Bayelein, chief of staff, German Africa Corps, 1941—42.
>
> General of the Air Force (General der Flieger) Paul Deichman, chief of staff of the German Second Air Force.
>
> Major Helmut Hudel, commander, 1st Battalion, 7th Armored Regiment, Tunisia.
>
> General of the Army (Generalfeldmarschall) Albert Kesselring, Commander in Chief, South, 1942—43.
>
> Regierungsbaurat (official title in the construction engineering profession) Dr. Sigismund Kienow, military geologist, German Africa Corps, 1941—43.

Brigadier General (Generalmajor) Gerhard Mueller, commander, 5th Panzer Regiment, 1942.

Lieutenant General (General der Kavallerie) Siegfried Westphal, who functioned in North Africa in 1941—43 as operation officer of Panzer Group (later Panzer Army Africa); chief of staff, German-Italian Panzer Army in Africa; commander, 164th Light Africa Division; chief, Operations Branch, German commander in Chief, South, attached to the Commando Supremo; chief of staff, Commander in Chief, South.

Dr. Wilhelm Wagner, medical officer, 21st Panzer Division, 1941—42.

Major Hubert Ziessler, commander of an artillery regiment, 1941—43.

I. PRIOR PLANNING

1. Intelligence Planning

a. *Desert Terrain and Climate*

When the first German units were shipped to Africa in February 1941, the officers responsible for the operational planning had no data of any kind on the nature of the terrain and circumstances in the desert. The intelligence data furnished by the Italians was extremely meager, and the Italian maps were so inaccurate and so incomplete that they were used only for lack of something better. For this reason, the German command had to obtain all necessary information itself through reconnaissance. In the papers found in his estate, Field Marshal Rommel wrote:

> It has probably never happened before in modern warfare that an operation of this type was undertaken with so little preparation. On 11 February, I reported to General Garibaldi, the commander in chief of the Italian forces and informed him of my mission. Initially, he showed no enthusiasm for my plan to organize defense positions in the region of the Bay of Sirte as a first measure. Using the poor and inaccurate Italian map material, I then proceeded to explain to General Garibaldi my ideas as to approximately how the war in Tripolitania should be conducted. Garibaldi, who was unable to give me any precise information about the terrain that would be involved, advised me to reconnoiter the terrain between Tripoli and the Bay of Sirte personally, and said that I could not possibly have any idea of the enormous difficulties this theater of war presented. Around midday I took off aboard a Type He 111 plane to reconnoiter the combat area. We saw the field-type fortifications and the deep attack antitank ditch east of Tripoli and then flew over a wide belt of dunes which presented a good natural barrier before the fortifications of Tripoli and would prove difficult to cross with wheeled or track vehicles. Then we flew across the mountainous country between Taruna and Homs, which appeared hardly suitable for operations by armored units in contrast to the patches of level terrain between Homs and Misurate.
>
> Like a black band the Via Balbia road could be seen extending through the desolate country, in which no tree or shrub was visible as far as the eye could reach. We passed over Buerat, a small desert fort on the coast with barracks and a landing stage, and finally circled above the white houses of Sirte. Southeast and south of this locality we saw Italian troops in their positions. With the exception of the

salty swamps between Buerat and Sirte, which extended only a few kilometers southward, we found no features in any sector that would favor a defense, such as, for instance, a deep valley. This reconnaissance flight supported me in my plan to fortify Sirte and the terrain on either side of the coastal road and to concentrate the mobile units for mobile operations within the area of the defense sector in order to counterattack as soon as the enemy started an enveloping attack.

From the above, it will be seen that Rommel himself had to gather the information on the terrain and on the peculiarities of the desert that he required for the conduct of operations. It was only during a later stage that the so-called military-geographical description was made available to the Germans, which gave a general survey of the terrain but was based mainly on information gleaned from literary works and contained none of the detailed information required by the troops. This data was of only small military value.

The military geological unit attached to the German Africa Corps commenced a systematic assembling of data and methodical reconnoitering immediately after arrival. The English maps captured by the German troops proved an excellent help. The results of the methodical reconnaissance were consolidated in what might be called a traversability map and in reports, and these were made available to the command. These maps contained the following details:

- Terrain that could be traversed by any type of vehicle in all parts and in all directions.
- Terrain outside the Pistes* that was moderately or poorly suited for vehicular traffic.
- Terrain with many steep cliffs.
- Salty swamps and depressions that were impassable after rain.
- Sand dunes that were difficult for vehicular traffic.
- Information on plant growth.
- Broken terrain.
- Impassable cliffs.
- Cliffs that were less steep and that could be traversed in numerous places.
- Passes over the cliffs, with information as to whether they could be used by wheeled or only tracklaying vehicles.

*Tracks.

• Trails, with information as to their usability for wheeled or tracklaying vehicles.*

The military geological unit compiling these maps consisted of two geologists and ten auxiliaries. However, they were inadequately equipped so that it was only possible to reconnoiter the areas that happened to be tactically important at any given time. Occasional inaccuracies and deviations in the lines marking the limits of the traversable terrain on the maps were unavoidable.

Here, a word might be said about the work of the British Long Range Desert Group that, apart from its intelligence and sabotage missions, carried out reconnaissance far behind the Italo-German fronts in Libya. The results obtained in this reconnaissance work formed the basis for the British maps on the Italian colony of Libya, which were incomparably better, so far as quality, accuracy, and detail were concerned, than the Italian maps. The British maps were considered a particularly valuable prize when captured.

b. *Scope of the Evaluation*

The above serves to show that in deserts, the command must employ adequate personnel with adequate equipment organized in specialized units if it wishes to obtain usable maps within a brief space of time. After the winter of 1941, the traversability maps served as permanent data for the German command. The preparations for attacks and for defense positions were based on them.

c. *Influence of Intelligence on Planning*

The available intelligence information was so inadequate in the spring of 1941 that it influenced in no way the employment of the German forces. As previously stated, Field Marshal Rommel had to gather the necessary information on the terrain and on the characteristics of the desert. On the basis of this information, he performed his mission of halting the British advance and preventing the loss of the whole of Libya.

d. *Availability and Evaluation of Terrain Intelligence*

The pamphlets *Military Geographical Descriptions for Libya, Northeast Africa, and Egypt* were published by the Military-Geographical Branch of the German Army High Command. Since they contained only information on cities, roads, oases,

---

*Some of these maps are found in Toppe's original, complete manuscript. See page ii.

and a general survey of the entire region, they could serve the command only as a source of general orientation, for which purpose they proved valuable. They contained very few important tactical details. They were put out in such large numbers that they could be made available down to regimental staff level. At these lower levels, their value was naturally restricted.

   e. *Use of Historical Data for Planning Purposes*

With the exception of the experience gained by General Graziani's army during its advance on Egypt in the winter of 1940, no information taken from military history was used in planning the campaign. One lesson that this experience pointed out is that troops that are not motorized are valueless in desert warfare and can do nothing whatever against a motorized enemy. General Graziani's army consisted almost exclusively of infantry units, and it was tied down, enveloped, and destroyed by the well-motorized British forces because it was unable to conduct mobile operations.

The African campaign took on such entirely new forms owing to the almost exclusive use of mobile troops by both sides in the desert. It was not possible in planning to make use of any examples taken from military history. Indeed, the methods of modern desert warfare were created by Field Marshal Rommel.

2. Operational Planning

   a. *General*

Prior to World War II, not a soul in the German armed forces imagined the possibility of it becoming necessary in any future war to conduct land warfare outside of Europe. This is why no particular attention was paid in the army to the military experiences of this type gained during World War I, particularly in the former German colony, German East Africa. It was only in 1935 that a subsection for colonial affairs was created in the Foreign Affairs Branch of the Reich Ministry of War. This subsection was staffed with only one officer who had fought in German Southwest Africa.

Prior to the outbreak of war in 1939, no preparations of any sort had been made in the German Army for any desert warfare that might possibly become necessary in the future. All preparatory work in the operational, organizational, and training fields had been restricted exclusively to preparations for the conduct of war on the continent of Europe. This was why a suggestion submitted by the Mapping and Survey Branch of the German Army General Staff in 1938 that the maps to be issued in the eventuality of mobilization should include maps of Denmark,

Norway, and North Africa was disapproved as entirely unnecessary by the appropriate representative of the Operational Branch under instructions from the chief of that branch.

It is an actual fact that early in 1941, the German troops reached the African theater of operations almost entirely unprepared for their new missions.

Up to the summer of 1940, the information available to the German Army General Staff on North Africa was restricted to the reports furnished by the German military attaché in Rome and reports from agents of the German counterintelligence service. From the autumn of 1940 on, Special Detachment Dora, a detachment of the German counterintelligence branch, was in Libya. Its main mission was to keep the French territories in Africa under observation. Most of the data on which the German military attaché in Rome based his reports came from his liaison officer attached to the governor general, who was simultaneously commander in chief of all forces in Italian North Africa, and on personal impressions gained while traveling. All positive information of a military nature on North Africa was taken from the manuals of the Foreign Armies Intelligence Branch (West) on the British, French, and Italian armed forces.

Originally, Hitler had decided to leave Italian dictator Benito Mussolini an entirely free hand in conducting operations in the Mediterranean theater, which was another reason for the small interest of the German General Staff in this subject. A change in this fundamental view of Hitler only took place in the summer of 1940, when it became evident, on the one hand, that Italy was apparently avoiding any decisive action in the Mediterranean theater, while the British, on the other hand, were continually reinforcing their troops in Egypt without their transportation being appreciably affected by the Italian Navy. At the meeting between Hitler and Mussolini in October 1940, the dispatch of a German panzer corps to Libya was discussed, but no decision was reached. Following the discussion, a general of the armored force who was attached to the German Army High Command was sent to Italian North Africa for an on-the-spot study of the possibilities of employing a German expeditionary corps there. Shortly after this, Italy rejected the support offered by Germany; quite obviously, Mussolini did not want any German military support in North Africa. The 3d Panzer Division, which in peacetime was garrisoned in the Berlin area, had been reorganized in all haste for employment in the tropics as a precautionary measure; it was now available for other employment. Later, when the British offensive, which gained huge initial successes, threatened to develop into a catastrophe for the Italian forces, Italy requested the dispatch of German forces to Libya.

The first unit to be transferred was the X Air Corps, which was sent to Sicily. So far as ground forces were concerned, the original plan was to send only a defense unit of brigade strength that was to be specially organized for the purpose, but it soon became evident that such a weak unit would not be able to give Germany's ally any really effective support. In January 1941, Hitler therefore decided to make a special corps of two divisions available—the German Africa Corps.

Meanwhile, a special staff for tropical warfare (*Sonderstab Tropen*) had been formed at the headquarters of the commander of the Replacement Training Army in Berlin. It was composed of officers who had fought in the German colonies in World War I and was to assemble as speedily as possible all experience that could be helpful in the training, organization, equipment, and employment of troops in desert warfare. However, the march of events was too fast so that the first units of the German Africa Corps landed in Africa when the staff had just commenced its work in Libya.

What has been said above goes to show that the German Army High Command was taken almost completely by surprise when the necessity arose to dispatch troops for warfare in the desert. In any event, the command had no time to make thorough preparations for this type of combat employment. For this reason, all preparatory work that was possible in the short space of time available had to be restricted mainly to the following measures:

(1) Medical examinations of all troops to determine their fitness for service in the tropics, with the application of very severe standards.

(2) Equipment of all soldiers with tropical clothing.

(3) Adaptation of a training program for combat in open terrain.

(4) Camouflage of all vehicles with a coat of desert-colored paint.

(5) Organization of special units to handle water-supply problems.

(6) Familiarization of the troops with the hygienic measures necessary in tropical climates.

(7) Orientation of the troops on the military-geographical conditions of the new theater of war and on the peculiarities of Germany's allies and enemies. In this respect, it must be mentioned that initially only one military-geographical bulletin was available. It had been prepared in a hurry and was not accurate in all points. A manual of instructions for the tropics was being drafted in the summer of 1942.

It was not possible within Germany to accustom the troops to the intense heat to which they would be exposed, particularly at that time of the year, the winter of 1940. To a certain extent, the troops that had to wait any length of time in Italy for transportation to North Africa adapted themselves automatically to the heat.

b. *Changes in Troop Organization and Equipment*

The composition of the units employed in Africa was the same as those in Europe. The pressure of time alone made any reorganization impossible in 1941, and later experience showed that no specialized organization is necessary for divisions and other units that are to be employed in desert warfare. However, it is necessary to have a far higher ratio of tanks and antitank weapons, since these are the two decisive weapons in the desert. It goes without saying that all units employed in desert warfare must be motorized.

The following special units were newly activated for employment in the desert:

(1) Water-supply companies, under the command of engineer officers. They were assigned to the corps and operated under the Water-Supply Branch of the corps supply and administration officer. These companies had pumps and equipment for the drilling of deep wells, while some of them had installations for the distillation of water.

(2) Water-supply transportation columns that were organized in the same way as ordinary supply-transportation columns but were employed solely in the transportation of water to the troops. They had no tank trucks or tank trailers as was customary with the British units but had to transport water in twenty-liter cans. This method of transportation proved extremely tiresome, quite apart from the considerable loading space required, which imposed an extra strain on the gas-supply services.

(3) Astronomical observation teams, directed by professional astronomers who were awarded regular or assimilated officer rank. These teams worked under the special staff officer

for surveying attached to the operations officer of the army, and their function was to establish geographical points by astronomical means. They were rarely employed, since no serious orientation difficulties arose because most of the fighting took place in the coastal region and not in the desert proper.

The following changes proved necessary so far as equipment was concerned: long-range artillery, long-range antitank guns, and tank guns decisively influence the course of battle in desert warfare, and it was therefore necessary to employ more long-range weapons. No alterations of the weapons themselves were necessary.

In their 87.6-mm guns, the British had a light artillery piece with a longer range than the Germans' guns, but the German forces in Africa soon received 100-mm and 170-mm guns that had a longer range than any of the British guns. In 1941, the guns of the German Type III tanks had a longer range than the guns of the British tanks, and this was the reason for the success of the German tanks in that year, but from May 1942 on, the British employed American tanks of the Grant, Lee, and Sherman types that mounted guns with a considerably superior range of fire. In the Battle of Gazala, these guns came as a disconcerting surprise for the German tank units, and in the first phase of the battle, the British were able to gain considerable successes.

Clothing and uniforms were entirely different from the clothing and uniforms worn in Europe. The German army uniform was made from a watertight linen, cut in a style approximating the traditional uniforms of the former German colonial defense forces. These uniforms proved unsuitable both in style and material. The material was too stiff and did not give adequate protection against heat or cold. In the early mornings, the material absorbed moisture from the dew so that it became intolerable to wear the uniforms. The British tropical uniforms, in contrast, were made of pure wool and were excellent. Large quantities of the British uniforms were captured and worn by the troops of the German Africa Corps (with the addition of German insignias). The Germans especially liked the British trousers. The tropical uniforms of the German Air Force, however, were good. Their color, a yellowish-brown, was more appropriate than other German uniforms, and they were made from a material that was of a lighter and better quality which was cut in a more appropriate style. Uniforms of olive-drab color proved unfavorable. In view of the normal camouflage difficulties in the desert, a yellowish-brown, which would have been a protective color, would have been best. High boots were unsuitable in every respect, since in hot climates, everything must be done to prevent soldiers wearing any apparel on the legs that restricts the circulation of the blood. In this matter, the troops helped themselves by wearing only slacks, most of which came from captured British depots and which the troops wore over their boots. The German shoe with laces and a cloth tongue proved suitable. The shorts issued to the troops could not be worn during combat, since they left bare legs exposed to injury by thorns and stones, and these injuries healed very slowly. The olive-drab caps with wide visors were excellent; the visor, in particular, was indispensable for the infantryman and for the gunner as protection against the intense glare of the sun. The tropical helmets that were issued could be used only in the rear areas and were entirely useless in combat. The German troops wore no steel helmets, in contrast to the British troops, whose steel helmets were more appropriate both in shape and weight, being lighter than the German helmets. The tropical coats issued, which were made from a thick woolen material, were good, but the English ones, which were fur-lined and reached only to the knees, were better. Owing to the stiff material from which it was made, the German tropical shirts were inferior to the British ones, which were made of so-called "Tropic" material. To protect the abdominal area of the body against the cold, the wearing of bellybands was obligatory, which proved a wise measure. Tropical helmets and mosquito nets proved an unnecessary expenditure. The majority of the troops got rid of them immediately after debarking from the ships, since they were not able to take them along owing to insufficient transportation space. The troops were also furnished wall tents, which had a special sun apron. With the exception of footwear, no leather was used in any article of apparel; it was replaced everywhere by thick linen.

The types of vehicles used were the same as those used in Europe. Vehicles with diesel engines were not used to avoid the necessity of transporting two different types of fuel. However, experience showed that it would have been advisable to accept this disadvantage in order to facilitate transportation, since fuel oil could have been transported in bulk containers, such as tank trailers. The excellent coastal road would have allowed the use of such transportation.

Volkswagens were used in great numbers and proved excellent. For use under desert conditions, the following alterations were made to adapt the standard model: air intakes were placed inside the vehicles to reduce the amount of dust taken in by the motor; in place of the standard tires, aircraft over-sized tires were used, which proved exceptionally good on rocky terrain and in sandy stretches. Because of their low air pressure, such tires reduced the shocks on rocky ground; while on sandy tracks, the wide treads of the tires prevented the vehicles from sinking into the sand and getting stuck. On the whole, however, the British motor vehicles, as a result of the extensive experience of the British in desert conditions, were superior to those of the Germans, being better adapted to the special conditions in respect to tires, power, higher ground clearance, and lower bodies. Double tires proved unsuitable, particularly in areas where the surface was covered with stones, as the stones became compressed in large quantities in the spaces between the tires. In the desert, motor vehicles must always carry something or other, such as rope ladders or grids, for the men to place underneath the wheels if they get stuck in the sand.

To reduce the effects of sand and heat, additional air filters for all types of vehicles were developed and used. They proved very valuable, although it was not possible to eliminate the effects of sand on the motors altogether.

Troops employed under desert conditions should be furnished a certain number of aircraft compasses, which should be mounted on the windshield next to the driver's seat. By means of a small magnet, deviations were excluded so that the driver was able to drive in the direction ordered. The sun compasses, which were developed for the same purpose, did not meet requirements, since they were too complicated and failed to function properly around midday, between 1000 and 1400. Pocket compasses were indispensable and had to be issued to each man individually, since the individual soldier plays a greater role in the desert than in any other theater of operations. The compasses used by the British, in which the dials floated on oil, were better than the German ones—and were preferred by the German troops when they managed to capture any.

The Germans failed to develop anything special as a protection against flies and other insects, which became particularly pestiferous in summer. Insecticides similar to FLIT were an urgent requirement for the combat units.

c. *Special Training*

It was not possible to give the troops, which were rushed to Africa suddenly and at short notice, any specialized training.

All that was done was to have them attend a number of lectures by specialists in tropical medicine and by officers who had a vague knowledge of conditions from previous travels. However, these lectures gave the troops wrong impressions of what they were to expect from the effects of heat, sand, insects, and diseases—instead of orienting them properly. The instructions on hygiene in the tropics, on the other hand, were good. Even units that were transferred to Africa during the further course of the campaign there received no real specialized training because the orders for their transfer usually came so unexpectedly that there was no time for this purpose. However, in a suggestion submitted to the German Army High Command by the army in Africa, the following training subjects were considered important:

(1) Exercises of all types in marching and combat in open, sandy terrain.

(2) Cover and camouflage in open terrain.

(3) Aiming and firing of all weapons in open terrain and at extremely long ranges.

(4) Recognition and designation of targets without instruments. The aiming and firing exercises were to be carried out by daylight, at night, in the glaring sun, during twilight, facing the sun, with the back to the sun, with the sun shining from one side, by moonlight, and with artificial lighting.

(5) Exercises during extreme heat.

(6) Exercises of long duration with no billeting accommodations.

(7) The construction of shelters in sandy terrain.

(8) Practice in night driving and in driving over sandy terrain.

(9) Night marching in level terrain.

(10) Orientation by compass, by the stars, and so forth.

(11) Driving by march compass.

(12) Recovery of tanks and other vehicles in sandy terrain.

(13) Laying and removing of mines in sandy terrain.

(14) Exercises in mobile warfare.

If it had been possible to train the troops in these subjects and to prepare them thoroughly, considerable losses could probably have been averted.

d. *Acclimatization of the Troops*

So far as the first divisions transferred to Africa were concerned, no measures were taken to accustom the troops to excessive heat. Some of the replacements sent forward later had the opportunity of spending a certain period in southern Italy or in the Balkans for acclimatization. The climate in these two regions is very similar to the climate in the coastal areas of North Africa. In the light of experience, however, a familiarization period is not considered absolutely essential, since the troops employed without a prior period of acclimatization proved no less efficient in combat than those who had lived for a time in southern Italy or in the Balkans. It was not the climate alone that caused the heavy losses that were suffered but the poor food, the hardships during combat, combined with the effects of the climate; the troops had in no way been prepared for these circumstances.

It proved very unwise to transfer units or replacements to the desert in summer, during the hottest part of the year and the time when the flies proved most troublesome. A parachute brigade provides a typical example. The brigade was transferred from Europe in July 1942, the hottest time of the year, and employed in defense in the rocky wilderness around El Alamein. The unit consisted of handpicked men, and within a very short while, more than 50 percent of them were sick from the combined effects of the heat, with its accompanying discomforts, and the inadequate diet. Shortly after the unit was committed, numerous cases of metabolic disorders set in, such as dysentery, jaundice, and festering sores that healed only very slowly. The causes were the brackish drinking water, which contained as much as one gram of salt per liter, and the inadequate diet, which consisted almost exclusively of canned foods. Blond and redheaded men with blue eyes and fair skins were particularly susceptible, while the brown and dark-haired types soon recovered from the disorders that were almost inevitable in the beginning. These points were not taken into account in earlier medical examinations—the main emphasis being placed on sound teeth and a strong heart. The result was that the elite units, such as the paratroopers, suffered particularly heavy losses. Even prior acclimatization would not have protected them.

The following experience was gained in respect to the acclimatization of persons to hot climates: men who had lived before in temperate zones stood the intense heat very well in the first year, during which they were far more efficient than the indigenous population and Europeans who had been living in the country for a long time. This proved to be the case when German troops were employed in Sicily, for instance, where summer temperatures are the same as those in the deserts of Africa. In the average case, however, the powers of resistance of the new arrivals declined after the first year, and their efficiency sank below the level of that of persons who had spent a longer time in the country. The newcomers' efficiency only started to improve gradually after a few years but never reached the same standard as that they experienced the first year. The following inferences can be drawn from this experience:

(1) No prior, lengthy acclimatization should take place for troops, since this would waste part of their first year of maximum efficiency.

(2) Only a brief transitional period should be allowed in a hot climate, during which the troops can be instructed in the manner of living under tropical and desert conditions and the best protective measures they should take without the added difficulty of enemy action.

(3) After approximately one year on active service in a hot climate, the troops should be rotated to some other theater of operations. The disadvantage that the experience gained by the men can only be exploited for a relatively short time must be accepted.

e. *Development of Special Equipment*

The following special types of equipment were developed:

(1) Special tropical clothing and uniforms, as dealt with in detail in section 2, b.

(2) Special air filters for motor vehicles, including tanks. This subject has also been discussed in section 2, b.

(3) Special medical equipment for use in tropical climates.

3. Logistical Planning

Logistical planning is an integral part of operational planning. In this operation, plans for the supply services also had to be prepared at top speed. The main concern in these plans was to provide transportation of supplies for the German troops by rail to Italian ports and by German or Italian ships to ports in North Africa. The selection of transportation media and supervision of the loading was the responsibility of a special branch, the Branch for Transportation to Africa. It operated under the command of the German military attaché in Rome. Unloading in African ports and further transportation of supplies to the troops was the responsibility of the supply and administration officer of the Africa Corps, later of the chief supply and administration officer of the *Panzergruppe Afrika*,* which later again was redesignated the Panzer Army of Africa and finally the German-Italian Panzer Army. Initially, all bulk commodities, as well as all troops, were transported by sea, but when shipping losses mounted, personnel were transported by plane.

In November 1941, Field Marshal Kesselring arrived in Italy as commander of the Second Air Force. In coordinated action with the Italian Navy and Air Force, his mission was to protect Germans and to prevent British transportation in the Mediterranean. It is said that shortly after his arrival, he sighed: "Now it is clear to me that in conducting a war across the sea, the proper delivery of the means of combat at their proper place is of far more importance than any worries as to whether the enemy should be attacked on the right or left flank."

It was not possible with the means available to the supply command or with any improvised measures to secure adequate supply services for the armored forces in Africa. To keep open the supply lanes or to open these lanes was the responsibility of the operational command, which rested with the Italian Supreme Command. The Wehrmacht High Command had supported the Italian Supreme Command but had also occasionally interfered in the conduct of operations. It was imperative that this problem be solved if an adequate supply service was to be secured for the troops in Africa. As no solution was found, the supply service collapsed as a natural consequence after all improvised means had failed. The following dates and information concerning the functioning of the supply services has been furnished by the German general attached to the Italian Supreme Command during the period from February 1941 to May 1943.

a. *February—May 1941*

The transportation of troops and supplies across the Mediterranean functioned without interruption. The convoys reached Tripoli regularly and almost without losses. Immediately after its capture, Benghasi was used as a port of debarkation. At the request of the German command, Italian submarines were used as early as April 1941 to transport fuel for the most advanced elements of the Africa Corps. They discharged their cargo at Derna. Coastal shipping along the African coast was organized with small ships and sailing boats with auxiliary motors.

---

*An armored army with no rear-zone administrative responsibilities.

b. *June—December 1941*

British surface and submarine craft interfered with the transportation of German troops and supplies. The losses in shipping space and in materiel were considerable. To relieve the situation, air transportation groups were employed to move troops and materiel, while naval barges transported tanks and important spare parts. The use of Bardia as a port of debarkation close to the front was prevented by the British Air Force. In December, Italian battleships had to be used to protect the convoys.

c. *January—June 1942*

During this period, transportation was favored by German superiority in the air, which was gained by the German Second Air Force under Kesselring and also by the fact that Malta was suppressed. The transportation of troops and supplies functioned smoothly and with few losses. Enough supplies were moved forward to enable the German-Italian Army to launch an offensive with limited objectives that advanced as far as the borders of Egypt in May—June. In addition, adequate supplies were stockpiled for a period of six to eight weeks against the eventuality of the air forces and naval vessels being employed in an operation to capture Malta.

d. *July 1942—May 1943*

As a result of Rommel's advance into Egyptian territory after the capture of Tobruk (this advance was contrary to the plans of the Italian Supreme Command), the supplies deposited in the Benghasi and Tripoli areas for the front were practically useless, since the distances were too great for transportation of supplies on land and coastal shipping was prevented by the British. The German Second Air Force was compelled to transfer some of its units stationed in Sicily and southern Italy to Africa and Greece to support the Panzer Army of Africa, which was fighting desperately at El Alamein. As a result, the Luftwaffe was so heavily engaged that it was unable even to screen Malta. The British forces on Malta regained their strength and employed new types of bombers that were equipped with radar and had a wider radius of action. The British succeeded in bringing German convoy traffic to an almost complete standstill. The Italian battleships were in port at Tarent and La Spezia, unable to operate because of lack of fuel. Losses in materiel and fuel were so heavy that it was barely possible to obtain adequate supplies from Germany. The sea routes to Tripoli and Benghasi were completely severed. Air transportation from Crete now played the major role, but quite naturally, the volume was far too small to meet even the most urgent demands of the front. In addition, the Wehrmacht High Command moved an infantry division from Crete to Egypt. This division had no motorized vehicles whatever so that it became an added strain on the transportation and supply services in Africa.

After the occupation of Tunis, the distances across the sea were admittedly shorter. Nevertheless, in spite of the use of the military transport ships, which had been constructed meanwhile, and numerous ships of the smallest types, it was not possible to relieve the strained supply situation. Anglo-American power in the air was growing steadily, and transportation capacities were sinking from day to day. Even a temporary increase of the quantities transported by air to 1,000 tons failed to bring any relief. Once the German-Italian forces in Tunis were enveloped, the Anglo-American fighter planes had such complete mastery in the air, even over the Straits of Sicily, that it was hardly possible for even the smallest ships to reach Africa safely. Around 20 April, the German-Italian air transportation units were subjected to a crippling attack.

Thus, the point must be brought out that as a result of the gradually developing Anglo-American supremacy at sea and in the air in the Mediterranean, North Africa was cut off from Europe. The German-Italian forces operating in Africa, therefore, could not be adequately reinforced or supplied. This lack of any possibility of maintaining supply traffic was not due to any failure on the part of the German or Italian headquarters responsible for the movement of supplies but solely to the fact that the German-Italian operational command did not succeed in keeping the supply routes to Africa open. Any examination of the question why these routes were not kept open or could not be kept open is beyond the scope of this study.

Plans for supplying the troops in the desert had provided for adequate supply transportation space and also an additional water-supply service. Each division had the same transportation space; the same motor vehicle and weapons-maintenance units; and the administrative, medical, and military police units as a division in Europe—plus a water distilling company. The corps supply services included an additional, special water-supply company and filter and distilling units and geological teams.

II. OPERATIONS
4. General Description of the Zone of Operations

The zone of operations in the North African campaign in Libya and Egypt consisted of a strip of land, sometimes as much as sixty kilometers wide, bounded on one side by the coast and on the other by the desert interior.

The ground surface was either firm gravel, sand-covered gravel, or mixed sand and gravel. Within this entire zone, large parts of which were level plain, the desert could be traversed by all types of vehicles. The only exceptions were patches of deep sand and steep wadis—which could not always be ascertained from the map—and salty swamps, such as those at Marada, roughly forty kilometers south of Marsa el Brega. Natural sinuous defiles were formed at Derna and the Halfaya Pass at the border between Libya and Egypt. It was also possible to create defiles by the use of mines.

Undulating, steppe-like terrain predominated, which consisted of low mounds and long ridges, whose average height above the surrounding terrain was from four to twenty meters. At times, these ridges had gentle slopes, and at other times, they rose steeply from broad, level valleys in which there were no watercourses. The summits were naked rock covered with loose rocks of varying size, which made motor traffic difficult but not impossible. In the valleys, the rocky bottom was covered by a layer of dust or clay of varying thickness. In dry weather, this ground could be traversed without difficulty by vehicles with four-wheel drive that were capable of cross-country travel—but not without raising dense clouds of dust. The steppe-like terrain had patches of camel's thorn shrubs, around which the dust had blown to form small dunes. Traffic followed the broad paths, called *Trighs* or Pistes, which connected the few settlements and water holes. This terrain extended from the coast to a line roughly thirty or forty kilometers inland. The coast itself was fringed by a belt of dunes behind which was a zone of salt swamps, called *Sebchen*, which were usually dry. This coastal zone was frequently used as a bivouac area for troops, since it offered good opportunities for digging in tents and vehicles and had good water-supply facilities. The only parts of the coast where there were no dunes were the cliff sections at Tobruk, Bardia, and Sollum. There, the coastal sector was often intersected by deep wadis and was difficult to penetrate.

Toward the interior, the steppe-like zone gradually merged with the desert proper, which is practically devoid of any type of vegetation. On the whole, motoring was easier in the desert proper than in the steppe-like zone, although movement was rendered difficult in rugged areas. In the desert, instead of rocky surfaces, patches with a deep covering of sand were encountered that made rapid travel possible. Here, the valley floors were clay pans, as flat as tabletops, which were submerged in water during the rainy periods. Only at the foot of steep cliffs was a rocky bottom found or a soft, sandy bottom, where vehicles might easily sink. This soft sand also covered the beds of the numerous wadis, by which the steep faces of the ridges are broken so that it was often extremely difficult to surmount the obstacles presented, even by comparatively low, steep ridges.

Farther south, these patches of soft sand increased in size and seriously impeded operations by armored units. The dividing line between those parts of the desert in which mobility was good and those in which it was bad is in eastern Libya and western Egypt, between the 29th and 30th degrees latitude. South of the 29th degree latitude, the vast dune-covered expanses began, and to cross them was considered quite a sporting feat.

We can thus see that the area suitable for military operations was confined to the relatively narrow strip along the coast and the southern desert zone, which was more favorable for rapid movement on the whole than the northern, steppe-like zone (if the tarred coastal road is left out of consideration).

Within the zone described, the following types of terrain obstacles were to be found:

a. Mountain Ranges

Three mountain ranges played an important part in the war in Africa, namely:

(1) *The Cyrenaica Mountains.* At points, these mountains reach a height of 875 meters above sea level and intercept the moisture carried inland by the north wind. The heavier rainfall here is the reason why the chalky ground carries a growth of macchia in contrast to the desert or steppe-like areas. The mountains rise in two, high, steep terraces that can be traversed at only a few points and are intersected by numerous deep valleys, which make it impossible to conduct sizable operations except along roads. South of the topmost ridges, the mountains slope down gradually to the desert terrain, which is good for vehicular traffic. For this reason, the Cyrenaica region was

vulnerable to attack from the south—a fact that Rommel recognized at once during his attack in the spring of 1941. For this reason, he delivered his main attack against Mechili, a desert fort designed to protect the southern approaches to the Cyrenaica. The fact that it was so easy to bypass is the reason why the Cyrenaica was never held with any degree of determination by either side during the entire campaign, although it could be called a natural fortress. During every retreat, every effort was made to pass through this region as rapidly as possible to avoid being intercepted.

(2) *The Gebel Nefusa Mountains.* These mountains protruded like a barrier between the coastal plains of Tripoli and those of Misurata. South of Tripoli, they rose to a height of 700 meters above sea level, the first 300 meters of which were a gigantic cliff. In the southeast, they descended in a gradual slope. At Homs, in the northeast toward the sea, their height was less than 200 meters above sea level. In the central part, this mountain range was extremely rugged, and motorized troops could only pass along the roads. The southeast slope was covered with a deep layer of wind-blown sandy loess that made vehicular traffic difficult. From the north, this mountain range formed an impregnable fortress. From the southeast, however, it was vulnerable to attack in spite of the mountainous and intersected nature of its approaches, since the attacking forces could find favorable assembly areas in the foothills and could approach under cover to the proximity of defense positions. Possibilities for bypassing the area existed and were taken advantage of by the British in the attack in January 1943.

(3) *The Matmata Mountains.* These mountains, a range in south Tunisia, had a steep drop of 100 to 200 meters in the east. In the west, they sloped down gradually to a high plateau, which was sandy in parts, while in other areas, the ground was good for motor traffic so that it could be crossed by motorized columns in spite of occasional difficulties. The steep, cliff-like wall in the east and north was interrupted by numerous wadis, through some of which an ascent to the high plateau was possible.

The Matmata Mountains narrowed down the size of the coastal plains of southern Tunisia considerably so that it was possible to organize a defense line at the narrowest point, at Mareth. However, the steep mountainside was only a weak protection against flanking attacks, since it could be bypassed with little difficulty. Only if the German-Italian forces had been numerous enough to hold all passes and if they had had a mobile reserve available to repulse any enemy attempts at detouring the mountains would this range have constituted an important factor in the defense.

b. *Steep Terraces*

Most of the steep terraces in the steppe-like terrain were not high and followed a course parallel with the coast. Thus, they hardly interfered with troop movements. In the numerous caves, overhanging cliffs, and gorges, good opportunities could be found for troop shelters, for which purpose they were frequently used, since they were the most effective protection against air attacks that was to be found. Some of the steep terraces and other similar terrain features, however, became of outstanding importance, namely:

(1) *The Northern Rim of the Qattara Depression, on Which the Southern Flank of the El Alamein Line Was Based.* This rim towered about 300 meters above the floor of the depression, which was 80 meters below sea level. Within the sectors held by the German-Italian forces, there were only three points at which motor traffic was possible, and even there, difficulties were encountered because of the deep sand. Throughout the entire campaign, no better protection for a flank was ever found than in the El Alamein line.

(2) *The Steep Terrace at Sollum Between the Bardia-Capuzzo High Plateau and the Sollum Coastal Plain.* There were two roads with numerous serpentine curves across the terrace, one from the Via Balbia—the tarred coastal road—the other from the Halfaya Pass road. During the period of positional warfare in the summer of 1941, the terrace was within the combat area.

(3) *Large-Size Wadis.* These were found in the Cyrenaica region and in the eastern approaches to the Tripolitanian Mountain and extended as far as the Bay of Sirte. Usually the bed of a wadi consisted of a layer of soft sand; less frequently, the beds were salty swamps with a growth of camel's thorn. The banks were usually steep but not continuous, since they were cut by numerous intersecting wadis. On the whole, wadis could be considered as terrain obstacles—but as obstacles that could be overcome without difficulty unless obstinately defended.

During the German-Italian retreat from El Alamein to Tunis, only one defense position was based on a wadi, namely the Buerat line, which extended along the Zem-Zem wadi south of the Via Balbia. After careful deliberation, however, the line was developed east of the wadi to prevent an approach by the enemy under cover and not on the low-lying west bank, since the west bank was dominated by the higher opposite bank.

The Buerat line could be bypassed easily. It was therefore evacuated by the infantry before the attack began and held only for a short while in a delaying action by mobile units.

(4) *Dune Terrain.* Large sandy areas were found close to the coast, near larger wadis, and in the desert proper, where the ergs* present barriers that were impenetrable for traffic.

Big dunes along the coast that interfered with traffic were found around Agedabia, on the shores of the Bay of Sirte south of Misurata, and in the neighborhood of Tripoli, thus mostly in western Libya. These dunes seriously impeded traffic off the roads, and even the roads were affected, since the dunes shifted constantly. After severe storms, the roads became so deeply covered with sand that they had to be cleared. For this reason, a constant road-maintenance service was necessary where the roads crossed dune areas.

A large area of dunes was also found north of the El Fareh wadi, between El Agheila and Marada along the shores of the Bay of Sirte. These dunes protected the German Marsa el Brega position against flanking attacks and forced the British to make a wide detour through the region south of the El Fareh wadi, where vehicular traffic was possible.

The big dunes of the desert proper were all south of the zone of operations, and only a section of them along the border between Libya and Egypt played a role of some tactical importance, since they afforded protection for the south flank of the German Alamein positions. The dunes in the desert proper were not crescent shaped like the dunes along the coast but formed continuous ridges between four and fifty meters high that usually extended from north to south. A number of these ridges, driven by the wind, formed a labyrinthian confusion of dune ridges with completely encircled hollows in which the firm ground could be seen. This enormous ocean of dunes formed what might be called a collection of honeycomb dunes. To cross them, it was necessary to have the best cross-country vehicles available and to drive at top speed at the first dune, breaking through its crest, and on driving down the opposite slope, to gather speed for the next dune. While driving in this way, vehicles were enveloped in a dense cloud of dust that reduced visibility to practically nil. In this way, one to two kilometers might be

---

*Large areas of shifting sand dunes (translator).

covered per day. Serious losses in personnel and materiel were unavoidable.

The Great Eastern Erg, a similar large dune area, extended from south Tunisia to south Algeria, close to the western border of Libya. If adequate German manpower had been available to extend the Mareth position across the Matmata Mountains and Fort Le Boeuf to this dune area, the German flank would have been as well protected as was the case in the Alamein line.

(5) *Salt Swamps.* These swamps developed at those points where the water in the subsoil of the desert rose to the surface. Owing to the constant evaporation that takes place in the desert, the salts carried by the water were deposited, and the resultant brine formed either a lake or, when mixed with sand and clay, a patch of thick, tough mud on which salt-marsh vegetation could take root. Once a person was caught in a salt swamp, it was impossible for him to escape without help. Vehicles sunk in salt marshes could be recovered but only on terrain that was not too swampy. In really soft, swampy ground, the vehicle had to be pulled out by another vehicle, which was often extremely difficult and could only be done if the latter was on firm ground and had a strong engine. Most of the salt marshes were crossed by fords that were known to the natives. Many of the fords could carry vehicular traffic so that any salt marshes within a defensive position should always be kept under observation, and all fords crossing it must be carefully reconnoitered with the aid of native guides. Frequently, the salt marshes dried out completely and then presented no obstacle at all.

The biggest salt marsh in the Libyan and Egyptian deserts was the Qattara Depression, the surface of which was eighty meters below sea level. This depression and its northern rim protected the flank of the El Alamein line. The swamp itself was skirted by a zone of soft sand varying between one and two kilometers in width, on which a few vehicles could travel with some difficulty. All other ground outside of the actual swamp, but within the Qattara Depression, was also soft and difficult to cross with vehicles. The salt marsh that was within the German zone of operations in the Marsa el Brega line was considerably smaller. Nevertheless, in conjunction with the sandy patches and dune areas, it provided good protection against a frontal attack, in spite of the fact that it had numerous fords. The salt marshes of southern Tunisia, called Schotts, were of more importance. The Schott el Djerid was the terrain feature that led to the decision to construct the Gabes line, which served as a rear line for the Mareth line. In most parts, the Schott el

Djerid was considered an impassable obstacle, but its eastern part, the so-called El Fedjad Schott, had numerous good fords that could be crossed without difficulty by vehicles.

Both Benghasi and Tripoli had good ports with ample capacities for shipping and landing, for which reason the former was used as the main supply base. The capacities in the ports of Derna and Bardia, as well as the naval port of Tobruk, were much smaller.

There was no continuous railroad in Libya. The two railroads, each about thirty kilometers in length, in Tripolitania and in the Cyrenaica were of no military importance.

The only permanent signal communications system consisted of an open-wire telephone line, on poles, from Tripoli to Bardia. The distances spanned were extremely great, and the line made only limited communication traffic possible. Furthermore, it was frequently interrupted by the frequent air attacks against the Via Balbia.

The water-supply facilities along the Via Balbia were adequate. The water holes in the desert, usually with a small supply of brackish water, were generally known only to the natives and were not indicated on maps.

During the main part of the year, the air was very hot but dry, the hottest months being June, July, and August. The highest temperatures registered around midday were about 140 degrees Fahrenheit. At night, even in summer, temperatures dropped to about 5 degrees. In winter, from November to January, the nights were quite cold, temperatures dropping to around 5 degrees and rising again during the daytime to about 85 degrees. Rain fell only in winter but was then sometimes very heavy, starting suddenly and swamping extensive areas, sometimes stopping all traffic, even on roads, for protracted periods. The only other moisture was the heavy dew at daybreak and in the evenings.

The outstanding weather feature was the sandstorms, which are called ghiblis. These sandstorms recurred pretty regularly every four weeks in all seasons of the year. They usually lasted three days, and since they reduced visibility to nothing, they brought all operations by ground and air forces to a standstill. During these sandstorms, the range of vision was often reduced to less than three meters so that orientation was impossible.

Owing to the wind from the sea, the climate in the coastal regions was almost always healthy. In spite of the enormous number of flies, there were few cases of malaria. On the other hand, the troops proved extremely susceptible to jaundice and dysentery.

5. Order of Battle of Army and Luftwaffe Units

Army. The first units to be transferred to Africa between February and May 1941 were the corps headquarters of the Africa Corps and headquarters units (the corps signal battalion and several supply units), together with the 5th Light Division, which was later reorganized to form the 21st and 15th Panzer Divisions.

During the summer months, a number of so-called oasis companies, a few battalions, and some coastal batteries were moved in, with an Africa Division Headquarters to control them. In the autumn of 1941, these units were consolidated to form a division, later designated the 90th Light Africa Division.

Thus, the German combat troops in Africa at the end of 1941 consisted of two armored and one light division. The two armored divisions remained under the command of the German Africa Corps. In the summer of 1941, this corps and the other army units in Africa were placed under the command of the newly created *Panzergruppe Afrika.* On 21 January 1942, this headquarters was redesignated Headquarters, Panzer Army of Africa, which designation was changed again in the autumn of 1942 to Headquarters, German-Italian Panzer Army.

In the summer of 1942, the 164th Light Africa Division and the Parachute Instruction Brigade were transferred to Africa. As they were transported by plane and since the sea transportation capacities were steadily sinking, these units never received their vehicles. Thus, they remained nonmobile to a great extent—a fact that was to have very adverse effects on the withdrawal from El Alamein.

In 1942, about eighteen batteries—that were not included in any of the divisions and consisted of Army Headquarters batteries, coastal batteries, and new batteries of captured guns— were consolidated as Army Headquarters Artillery. This artillery was organized in two regiments and was placed under the command of the commander of artillery in Africa. In addition, the reconnaissance battalions of the 15th and 21st Panzer Divisions and the 580th Reconnaissance Battalion (a general headquarters [GHQ] unit) were consolidated to form a reconnaissance brigade under the immediate control of the Army Headquarters.

The army also had the 900th Engineer Battalion, formerly a GHQ unit, available as a headquarters unit.

At the end of 1942, therefore, the ground forces employed in combat consisted of the following:

    2 armored divisions
    2 light divisions
    1 parachute brigade
    1 reconnaissance brigade
    2 regiments of headquarters artillery
    1 engineer battalion
    The 288th Special Unit, a reinforced battalion originally organized as an elite battle group for employment in the Middle East

This list does not include the numerous units available to the army for logistical support.

The above divisions were organized as follows:

a. *Armored Divisions—*

    Divisional headquarters
    2 armored infantry regiments, each of 2 battalions
    1 tank regiment of 2 battalions, with a table of organization of 100 tanks each
    1 artillery regiment of 2 light battalions and 1 heavy battalion (9 batteries, with 24 light field howitzers, 8 heavy field howitzers, and four 100-mm guns)
    1 antitank battalion of 3 companies, each with 3 guns with prime movers
    1 engineer battalion of 2 companies
    1 signal battalion, with 1 telephone and 1 radio company
    Supply and transportation units
    Total strength of each panzer division, 12,000

b. *Light Division—*

    Divisional headquarters
    3 infantry regiments of 2 battalions each
    1 artillery regiment of 2 light and 1 heavy battalion (24 light and 12 heavy field howitzers)
    1 antitank battalion of 3 companies (armament as for panzer division)
    1 engineer battalion of 2 companies
    1 signal battalion (as for a panzer division)
    Supply and transportation units
    Total strength of the light division, 12,000

c. *Parachute Instruction Brigade—*
   Brigade headquarters
   4 battalions
   1 engineer company
   1 light artillery battalion
   1 mixed signal company
   Total strength of the Parachute Instruction Brigade, 5,000

In addition to the above, the following units were landed in Tunis and employed in combat from November 1942 to the end of the campaign:

   Considerable parts of the 10th Panzer Division
   1 battle group of regimental strength of the Hermann Göring Parachute Panzer Division
   Considerable parts of three infantry divisions, a number of GHQ armored battalions and the German Arab Legion, the latter of which was a unit of regimental strength

Thus, the ground forces employed in combat in the African theater of operations were equivalent to the following:

   3 armored divisions at full strength
   2 light divisions at full strength
   2 infantry divisions at full strength
   1 parachute brigade

Air Forces. The fact must be stressed at the outset that the air force units stationed in Africa were kept at a low level of strength to avoid further complicating the already difficult supply situation. Additional air support was given by air force units stationed at Italian or Greek air bases, which were transferred occasionally, for temporary periods, to Africa.

Three phases must be differentiated in respect to the organization and composition of air force units stationed in Africa, namely:

Phase A, February—November 1941

Phase B, December 1941—December 1942

Phase C, January—May 1943

Phase A

Command: German Air Force commander in Africa. The commander was subordinate to the X Air Corps (stationed in Athens and later on Crete) and was in tactical support of the

Africa Corps (later the Panzer Army of Africa).

Flying forces in Africa:

> 1 long-range reconnaissance squadron (F-121-type planes)
> 2 squadrons of the 14th Close Range Reconnaissance Group
> 1 fighter group, later replaced by the 77th Fighter Wing of 3 groups
> 2 groups of the 3d Dive-Bomber Wing
> 1 destroyer plane group
> 1 desert-rescue squadron

Antiaircraft artillery:

> 1 regiment of 4 battalions, tactically assigned to the Africa Corps (later *Panzergruppe Afrika*)

Air Signal troops:

> 1 air signal battalion

Logistical support troops:

> 1 team detailed by the Luftwaffe General in Italy

Phase B

Command: German Air Force commander in Africa. The commander was subordinate to the Second Air Force and was assigned tactical support of the *Panzergruppe Afrika* (later Panzer Army of Africa).

Flying units in Africa, as in section a above.

Antiaircraft artillery:

> Organized in the summer of 1942 to form the 19th Flak Division and tactically assigned to the Panzer Army
> Air Signal troops: as in section a above
> Logistical support troops:
> From 1942 on, Air Force Administrative Command, Africa, was controlled by the Luftwaffe General in Italy

Phase C

Command: Air Corps Africa, with Air Commanders 1 and 2. The Air Corps Africa was subordinate to the 2d Air Force and was required to cooperate as follows:

> Air Corps Africa with Army Group Africa
> Air Commander 1 with the Fifth Panzer Army
> Air Commander 2 with the German-Italian Panzer Army (later redesignated the Italian First Army)

Flying units in Africa:
  2 fighter wings (53d and 77th)
  1 dive-bomber wing of 2 groups
  1 destroyer plane group
  2 antitank plane squadrons
  Reconnaissance, units as in section a above.

Antiaircraft artillery:
  19th Flak Division, tactically assigned to the German-Italian Panzer Army (later redesignated the Italian First Army).
  20th Flak Division, tactically assigned to the Fifth Panzer Army

Air Signal troops:
  1 reinforced air signal battalion

Logistical support troops:
  Air Force Administration Headquarters Tunis, with three air base areas

The organization and the main items of armament as of January 1942 were the same as in Europe, with the exception of the additional supply units assigned for service in the desert, namely, the water supply units, the meteorological survey teams, and so forth. It must be emphasized in respect to the tables of organization that the units at no time had the stated authorized strengths. The actual strengths were constantly subject to fluctuations according to the losses suffered and the replacements received. Thus, the combat efficiency, which also depended on the shipment of replacements in personnel and materiel, also fluctuated.

6. Reasons for Changes in Organization and Equipment

Initially, the German units were transferred to Africa with their normal tables of organization and equipment. Changes that were effected immediately in respect to equipment were as follows:

   a. All vehicles were immediately fitted with new special dust filters.

   b. Special units, namely, water supply companies, water transportation columns, and geological teams, were organized immediately to take care of water supply and transportation problems. However, owing to the steadily increasing transportation difficulties, large portions of these units remained in Italy until the campaign was over.

c. All vehicles were camouflaged by a coat of desert-colored paint.

d. In reference to uniforms and other clothing, the troops were issued tropical shirts; khaki-colored linen jackets, breeches, and shorts; lace boots and shoes, both cotton-lined; linen caps with visors; tropical helmets; bellybands; and woolen overcoats.

In 1941, the following additional changes became necessary:

a. The antitank battalions arrived in Africa with 37-mm antitank guns. In the summer of 1941, these were exchanged for 50-mm guns, which were exchanged again in early 1942 for captured Russian 76.2-mm antitank guns. This was necessary because of the increased effectiveness of weapons used on both sides.

b. From early 1942 on, all infantry units were also assigned antitank guns, since tank warfare is the deciding factor in desert warfare, where the antitank gun becomes of even greater importance to the infantry than the machine gun. The aim of furnishing each battalion with eighteen 76.2-mm antitank guns was never achieved.

c. Types I and II tanks, some of which were armed with machine guns and some with 20-mm guns, were withdrawn after the summer of 1941 and replaced by Type III tanks, which had 50-mm guns. These, again, were replaced after the winter of 1941—42 by Type IV tanks, which had 75-mm guns.

d. All motorcycles were replaced by Volkswagens. Even the half-track motorcycles that were used for a while proved unsatisfactory.

7. Descriptions of the More Important Battles

a. *31 March—19 April 1941: The First Counterattack to Reconquer the Cyrenaica*

Contrary to the views of General Garibaldi, commander in chief of the Italian forces in Africa, Rommel, who had arrived in the theater of operations on 11 February 1941 as commander of the German Africa Corps, was of the opinion that waiting would worsen the situation. The British forces were still in a long drawn-out column and were momentarily in a precarious condition, which had to be exploited immediately. Rommel was able to substantiate his opinions by reconnaissance and, therefore, his views prevailed. Immediately after the 5th Light Division commenced landing at Tripoli on 11 February 1941 and moving up to the front, Rommel commenced a series of reconnaissance thrusts west of Agheila on 24 March, which he followed up on 31 March by an attack with limited objectives in the direction of Agedabia. The sole objective of this attack was to drive back the British troops in the advanced positions of Agedabia. Since these British troops retreated immediately, Agedabia itself was attacked and taken on 1 April, the enemy withdrawing toward Benghasi. The attack toward Benghasi that then followed was also successful, and on 4 April, that city and the port were taken by German forces.

In view of the obvious weakness of the British, who had been taken by surprise by the German attack, it seemed advisable to continue the advance. Rommel decided not to continue the pursuit through the Cyrenaica but rather to launch an enveloping attack through the desert in order, if possible, to prevent the retreat of considerable enemy forces. For this reason, he pushed forward the bulk of the 5th Light Division south of Benghasi straight through the desert towards Mechili and Derna, with weaker forces moving by way of Msus in a flanking drive. This move also succeeded, and on 6 April, more than 2,000 prisoners were taken at Mechili, Derna being captured on the same day.

On 9 April, the pursuing columns reached the Libyan-Egyptian border at Bardia so that all territory lost in Libya had been recovered. Only the Tobruk fortress remained in British hands. It was enveloped with weak forces by 11 April. Two attempts to take it in raids on 13 and 14 April and a third attempt, in a properly prepared attack on 30 April, failed. The forces available were inadequate for the task.

Rommel now had to decide whether to break off the siege of Tobruk and to withdraw to the elevated terrain of Ain el Gazala or to maintain the siege—with the disadvantage that he would have to establish a second front in a line level with Sidi-Omar-Sollum-Bardia. He decided on the second solution. Chiefly Italian troops—namely, the X and XXI Corps, with a total of four infantry divisions (which were to be increased to five at a later stage)—were to maintain the siege of Tobruk. The Sidi-Omar-Sollum front was held only in strongpoints in order to release the bulk of the German forces for mobile employment in the open field. To summarize:

(1) The units that took part in the actual offensive operations were as follows:

The 5th Light Division at that time consisted of three battalions; one tank regiment; and one each of reconnaissance, light artillery, antitank, engineer, and signal battalions.

(2) The important factors that brought about this speedy and thorough success were the following:

 (a) The momentary weakness of the British forces, whose supply transportation had not yet been able to catch up fully with the rapid advance.

 (b) German supremacy in the air.

 (c) The direct attack through the desert, which the enemy had not expected.

(3) A special feature of these operations was the advance through the desert from south of Benghasi toward Mechili and Derna, which was ordered by Rommel in spite of the serious misgivings of most of the commanders serving under him. The actions brought out the necessity of having the commanders of mobile units far ahead in the unit column in desert warfare and of employing all means, including liaison planes, to maintain contact within the pursuing force. There is no other possible way of remaining close on the heels of the retreating enemy.

(4) Logistical requirements were not given the proper consideration. This is the reason why some of the units failed in the desert. But, on the other hand, Rommel could not afford to wait for the arrival of further fuel transports, as he would then have lost contact with the enemy.

(5) Here, for the first time, the units had to cross a long stretch of desert, some units for 300 kilometers and more, and while doing so, they had to gather the experience they lacked. This experience included recognition of the necessity to carry along ample supplies of fuel and water and the difficulties of orientation. In the desert, it is almost impossible to establish one's position by the sun, since the sun is usually almost directly overhead. The available maps, which were reprints of Italian maps, were inadequate. Practically no reference points existed so that all orientation had to be done by compass. Furthermore, the eyes of the troops had to become accustomed to the glare of the sun, which made contours unclear. Thus, it was extremely difficult to recognize objects—for instance, to differentiate between tanks and trucks.

(6) Together with the fact that any movement caused immense clouds of dust, the above factor was originally exploited by Rommel, who had his supply and baggage trains move in tank formation in order to mislead the enemy. Later, this came

to the notice of the enemy, and later attempts to employ this ruse were unsuccessful.

(7) At that stage, the German forces suffered little from enemy air attacks.

(8) Here, for the first time, the 88-mm antiaircraft guns proved effective antitank weapons. Later, they became indispensable for this purpose.

b. *May—June 1941: Battles for the Positions on the Border*

The British left Rommel no peace and in these months seized the initiative several times in attempts to take from the Germans the border positions that commanded the outpost area. The British, particularly, attempted to take the Halfaya Pass. In the mountain range extending from the coast to the interior of the desert (a distance of more than thirty kilometers), the Halfaya Pass was the only point at which tanks could cross.

On 15 May, the British succeeded in recapturing Sollum, Capuzzo, and the Halfaya Pass. Two days later in an immediate counterattack, Rommel succeeded in retaking Sollum and Capuzzo, while the Halfaya Pass remained in British hands. On 27 May, however, the pass was finally retaken in an attack in which the 15th Panzer Division, which had meanwhile reached the front, also took part.

On 15 June, after careful preparations, the British launched a major offensive that aimed at retaking the border positions and advancing on Tobruk. The British bypassed the German border positions and pushed forward almost as far as Bardia. The situation was critical. However, on 17 June, Rommel, again employing the 15th Panzer Division, succeeded in defeating the enemy by concentrating his forces in an attack on the west flank of the enemy, who had advanced northwards. The enemy forces were compelled to withdraw southward to avoid the encirclement of some of their units.

The more important features of these operations are as follows:

(1) The pursuit phase was now over, and the actions described were those of attack and defense.

(2) Stronger forces were employed on both sides than had hitherto been engaged. On the German side, both divisions, the 5th Light and the 15th Panzer—minus certain elements tied down on the Tobruk front—were fully employed, as well as one Italian division. Without the 15th Panzer Division, the German

forces would not have been able to hold their own, particularly in the battle from the 15th to 17th of June.

(3) Whereas the fighting during the pursuit in March and April took place on either side of the Via Balbia, all the actions just described took part in the desert.

(4) The German side no longer had absolute mastery of the air; British bombing units were taking part in the fighting in concentrated attacks for the first time.

In these skirmishes and battles, the 15th Panzer Division gained its first experience in desert warfare. The fields in which experience was gained were the same as those described for the 5th Light Division in section 7, a.

New features in this operation were as follows:

(1) For the first time, all German units were exposed to lively enemy activity in the air, a feature they were to experience daily from now on. At first, several instances occurred where severe losses were suffered owing to the bunching up of vehicles and troops. It was weeks before the troops learned to counter this new combat factor by a wide dispersal of units in breadth and depth—a particularly important requirement in the desert, where no cover whatever is to be found. (The minimum distance between vehicles should be 50 and if possible 100 meters.) It also proved necessary to dig in immediately all vehicles that were halted for any considerable time. They were to be dug into the ground to at least a depth that protected the axles in order to lessen the effects of bomb fragments. In the same measure, it was also necessary to camouflage the vehicles. This was only possible with the use of camouflage nets so that it was extremely difficult. Furthermore, it was now necessary for each and every man to dig a foxhole as protection during air raids.

(2) The danger of radio stations being intercepted and located made it imperative to have all radio instruments, and particularly central radio stations, removed at least one kilometer from headquarters sites in order for them not to interfere with the functioning of staff headquarters. The resultant delay in the transmission of orders and reports had to be accepted as an unavoidable disadvantage. This delay had to be reduced as far as possible by the use of messengers with motor vehicles.

During the time discussed above, consolidated measures were also taken in the envelopment of Tobruk. The intention to withdraw all German troops from the besieging force could not be carried out, particularly at Ras el Medauuar, on the southern front, where two German battalions remained in position until the autumn of 1941.

(3) The danger of enemy tanks breaking through the front made it necessary to develop all-around defense positions protected by antitank mines. Rommel issued a bulletin describing the development of such positions, each held by a reinforced company in a system of strongpoints. Above all, this system was adopted along the border, where the Italian *Savona* Division was employed in addition to five German oasis companies.

c. *July—Mid-November: The Siege of Tobruk and Preparations for the Attack*

It was clear to Rommel that Tobruk had to be taken as soon as possible, and it was obvious that the enemy would do everything possible to prevent this happening. Speed was therefore necessary. The following factors made it difficult for Rommel to take the steps that he recognized as essential:

(1) The necessity of awaiting the arrival of further troops, infantry and particularly heavy artillery, and large supplies of ammunition from Europe, since the available forces were inadequate.

(2) The steadily decreasing capacities for German seaborne transportation as the result of the mounting losses of ships.

As early as July, it became evident that it would definitely not be possible for the Germans to commence any systematic attack before mid-September. At an early stage, it was realized that this deadline would have to be extended to October, then to November, and finally to December. Gradually, German doubts grew that the attack could be launched before the expected British offensive commenced.

The summer months were spent in executing the following measures:

(1) Reinforcement of the enveloping forces by artillery and through development of the terrain.

(2) Improvement of training.

(3) The movement of large quantities of ammunition and fuel to Benghasi and farther east.

(4) Improvement of the medical services, which had hitherto necessarily been neglected.

(5) Overhauling and maintenance of arms, equipment, and vehicles.

The following is pertinent to the activity during these months:

(1) All attempts to reduce frontage and thereby strengthen the enveloping line failed, since the Italian troops, by whom the greater part of the line was held, were not able to withstand counterattacks by the British.

(2) The reinforcement of the artillery forces was pressed forward vigorously; for this purpose, a special artillery commander was assigned. Flash and sound ranging proved indispensable in the location of the enemy batteries.

(3) Again and again, the order had to be stressed that all units employed were to dig themselves in as deep as possible in order to reduce losses.

(4) Demonstration exercises took place to improve the standard of training, with particular emphasis on combined infantry-artillery-tank, artillery-tank-air force actions and the practical application of the all-around-defense strongpoint system.

(5) It was only from Tripoli and Benghasi that ammunition and fuel supplies could be moved forward to the front. The lack of any rail connections proved a serious disadvantage. Investigations showed that to construct a railroad to meet even the most modest demands, at least 60,000 tons of shipping space for locomotives, cars, rails, understructures, and so forth, would be required, and a period of about twelve months for the Tripoli-Benghasi section and an additional three months for the extension to Derna would be needed.

Ammunition and fuel had to be stored in the open, both in the vicinity of the ports and near the front, since tank installations and shelters were nonexistent. This made wide dispersal and the burying and camouflage of all supplies at the storage depots all the more important. These precautions were frequently disregarded so that unnecessary losses occurred.

(6) Warm clothing after sundown was particularly important in the desert, and especially so for new arrivals, as a precaution against dysentery and skin diseases, since the difference between the daytime temperatures and those at night was extreme. After sunset, it was absolutely essential for every man to wear trousers and bellybands. Experience showed, in fact, that it was advisable to wear the latter day and night.

An appropriate diet was essential to prevent jaundice, which occurred frequently. A large proportion of the cases of jaundice that occurred in 1941 resulted because the rations issued included large quantities of pulses (i.e., leguminous plants) and conserved meat with a high fat content. Above all, food with a high Vitamin B and C content proved necessary, and on the whole, the food had to be light. Vitamin C tablets could not take the place of fresh vegetables. Owing to inadequate transportation space aboard aircraft, it was usually only possible to fly in fresh vegetables and fruit for air force personnel in Africa.

(7) In weapons maintenance, protection of the inside parts of the weapons against sand proved a particularly important point. For this reason, all bolts and moving parts of the weapons were wrapped in sailcloth—besides the use of the standard muzzle covers. All weapons had to be cleaned very carefully, but after cleaning, they were oiled only thinly; otherwise, the dust would eat its way into the surface. No special means to protect the weapons against dust were available. What has been said about the care of weapons applies in equal measure to the care of other equipment and motor vehicles.

(8) A high standard of training in the use and care of weapons, equipment, and vehicles was particularly important in desert warfare, and the work of the higher-echelon ordnance technicians in handling weapons, equipment, and vehicles was of great significance in maintaining the combat efficiency of the troops.

In an overseas theater of operations, extensive maintenance services with well-equipped workshops for the repair and maintenance of weapons, tanks, and other motor vehicles were just as indispensable as stocks of all types of spare parts, particularly for tanks.

It was also during the summer that Italian forces constructed the road to bypass Tobruk; it was roughly sixty kilometers long. This road was graveled and tarred, and its construction, which took three and a half months in the heat of summer, must be regarded as an outstanding performance. On the whole, the German troops, who were unaccustomed to the heat, also came through the summer with few losses.

On 14 and 15 September, Rommel launched a reconnaissance in force from the border positions in the direction of Bir el Habata (in the Egyptian desert). The operation was directed by the headquarters of the German Africa Corps and was carried out by the 21st Panzer Division, which had been organized from the 5th Light Division. This operation, which was designated *Sommern Achtstraum*, must be considered a failure, since it failed in its purpose of discovering how far the British were along in their preparations for their offensive. No opponent was contacted, as the British reconnaissance forces had recognized the German intentions and had withdrawn in good time. On the other hand, the 21st Panzer Division suffered considerable losses in a number of air attacks, owing to the fact that it lost three and a half hours on Egyptian terrain in refueling, as the fuel trucks first had to be moved forward. These losses could have been avoided if sufficient fuel had been carried along in cans and if the fuel column had accompanied the combat units. Further losses were sustained while moving back through German minefields, the locality of which was not known precisely to the various units.

d. *Mid-November 1941—Mid-January 1942: Repelling the British Autumn Offensive and the Retreat to the Gulf of Sirte*

The British offensive opened on 18 November 1941. At the strategic level, it had been expected, but nevertheless it came as a tactical surprise. This was because, from the end of October on, the German air reconnaissance hardly ever succeeded in penetrating into Egypt, and the enemy had concealed all general preparations and signal traffic with extreme skill.

Excluding the Tobruk garrison (one and one-half divisions and one armored brigade), the ground forces of the enemy, which had meanwhile been consolidated to form the British Eighth Army, consisted of XIII and XXX Corps headquarters, 3 motorized divisions, 1 armored division, and 1 armored brigade—with a total of about 700 tanks.

Apart from the Italian 5th Division, one German division, and the GHQ artillery besieging Tobruk, Rommel had available for operational employment: 2 German armored divisions, with roughly 360 serviceable tanks; 1 Italian armored division, with roughly 150 inferior tanks; and 1 Italian motorized division, the efficiency of which was also limited.

The XXX British Corps, with the bulk of the available armor, advanced through Maddalena to relieve Tobruk, while the British XIII Corps enveloped the border positions from the south.

The 21st Panzer Division, which was echeloned forward in the direction of Bir el Gubi, had the mission of halting the British advance but met with no success in its efforts. For a long while, the situation remained unclear to Rommel, because the division reported too infrequently and its reports were confusing. On 23 November, it seemed that the situation would improve when Rommel succeeded at Sidi Rezegh in battering the British XXX Corps so badly that the commander of the British Eighth Army seriously considered breaking off the offensive. Overestimating the scope of his success, Rommel then decided on an enveloping pursuit on the next day. On 24 November, he advanced with the Africa Corps in the direction of Maddalena, then wheeled north, and arrived back at the Tobruk front on 28 November. Here, the situation had developed unfavorably in the meantime, since the enveloping forces had not been able, in the long run, to beat off the repeated attempts of the enveloped British forces to fight their way out. The enveloping ring had been breached already on 22 November at El Duda, although the breach was locally restricted. The Africa Corps now only had roughly 100 serviceable tanks available and was no longer strong enough to restore the situation. Thus, it became necessary to raise the siege on 7 December. The difficult maneuver of swinging the Italian Division, the Africa Division, and the artillery forces westward was performed successfully, and a new front was established in the Ain el Gazala line. This position had to be abandoned on 16 December because it was in danger of being enveloped from the south.

For tactical reasons, Rommel thought it impossible to hold the Cyrenaica, which protruded northward and provided ideal opportunities for the enemy to bypass it, although the Italian command, for political reasons, demanded that he do so. He therefore decided to withdraw toward Benghasi-Agedabia.

This movement was carried out in the following manner:

(1) The Africa Division was dispatched through the Cyrenaica in order to take possession of the important town of Agedabia before the arrival there of an enemy column reported to be advancing westward through the desert.

(2) The Italian Division was also moved through the Cyrenaica to the rear on vehicles of the supply transportation columns.

(3) The Africa Corps and the motorized Italian division at Mechili were to advance straight through the desert to Benghasi. The motorized units carried out the movement successfully, but the available transportation space was unfortunately inadequate to move all Italian infantry forces to the rear.

On Christmas, the Panzer Group was ahead of Agedabia. On the last day of the year, the Africa Corps, which was echeloned to the right, was once again clearly successful in a defensive action against the pursuing enemy forces and destroyed a large number of enemy tanks.

Two additional factors alleviated the situation for the armored group. One factor was the considerable reinforcement of the German air forces through the transfer of the Second Air Force Command—with the II Air Corps from the Eastern Front—to Italy and Sicily, a transfer that had commenced toward the end of November. As a result of this closer disposition of German air forces, the hitherto overwhelming superiority of the British in the air was somewhat reduced. The second factor favoring the Germans was that the extremely tense supply situation was relieved by the arrival of two big convoys at Tripoli with supplies of all sorts, replacement tanks, and two tank companies and artillery, which became organic to the units in Africa. This was the first supply shipment to arrive between 16 September and 15 December 1941, during which period not a single ship had reached African ports.

In spite of the relieved situation, Rommel decided not to await the enemy attack in the Agedabia area, and in early January retired to the Marada-Marsa el Brega line, where he hoped that his right flank would be better protected by the salt marshes.

The more important lessons learned in the battle (that has been described above in broad outline) are as follows:

(1) The old maxim that reports should be sent in as frequently as possible was frequently not observed, although orders had been given that a brief radio report was to be sent in every two hours—with the provision that the single word "unchanged" or a statement of position would be sufficient.

(2) Similarly, not all of the units reacted automatically to any development by carrying out new reconnaissance.

(3) The use of the "directional line," with the aid of a few natural reference points in reporting and in issuing orders, proved an excellent system, particularly under desert conditions. This system is as follows:

A directional line is drawn between two points on the map, from point A to point B. Starting at point A, this line is marked and numbered consecutively at intervals of one centimeter. Positions can now be reported by this line; for instance, three right of thirty-seven would indicate a point on the map three centimeters east of thirty-seven on the map. The starting number for the consecutive numbering of the centimeter markings can be fixed as desired. Brief orders can be signaled in clear text with the aid of the directional line. It goes without saying that the line must be changed frequently.

(4) Not all unit commanders or their general staff officers were at all times precisely informed on the supply situation of their units, which had adverse effects on operations. At all times, all unit commanders and their assistants must know exactly how much of the more important types of fuel and ammunition their units have available, what quantities of supplies are to be expected within the next twenty-four hours, and what percentage of the most important weapons are ready for action. This knowledge is indispensable as a basis for all command decisions.

(5) Under desert conditions, the frequent penetrations by armored forces and the open terrain expose the higher-level staffs to danger to a far greater extent than is the case in any other theater of operations. Thus, all staffs must be protected by close-defense antitank weapons. For this reason, the *Panzergruppe* and the Africa Corps had organized so-called combat detachments consisting of tanks and antitank and antiaircraft guns in battalion and company strength. These units also proved useful as a tactical reserve.

(6) One feature peculiar to the desert operations in 1941—42 was the constant threat to the southern flank of the side that happened to be on the defensive—the northern flank generally being well protected since it extended to the coast. The danger to their right flank made it necessary for the Germans to have strong mobile forces, with ample supplies of fuel, echeloned far to the right to avoid their being forced to abandon a position bypassed by the enemy.

(7) In desert warfare, retrograde movements will usually be restricted to roads and will be difficult owing to the lack of natural obstacles favoring new defensive lines. Only if a firm control of units is maintained during retrograde movements over great distances will it be possible to prevent a retreat continuing beyond the intended point and endangering unit cohesion. For this reason, it is necessary to compel the rear echelons, such as transportation columns and so forth, to halt at intervals.

(8) Owing to the dust that is caused by any movement on the ground, it is difficult to differentiate between friend and foe from the air. For this reason, bombing "stoplines" must be established and clearly defined with due allowance for safety factors.

e. *Mid-January—End of May 1942: The Counteroffensive to Retake the Cyrenaica and Preparations for the Attack on Tobruk*

On 10 January 1942, the Panzer Group reached the Marada-Marsa el Brega line, where new defense positions were to be established. However, the remaining units, particularly the Italian troops, had been so far reduced in numbers during the previous fighting that they would hardly be able to hold the sixty kilometers of frontage against any major attack by the enemy for longer than twenty-four hours. A careful examination of the situation revealed that the enemy forces were still echeloned far to the rear so that they were in a critical situation similar to that which they had been in during the previous year. The coastal road remained closed to them until 17 January, when Rommel approved the surrender of the troops holding the Halfaya Pass since their supplies of ammunition and foods were exhausted. A careful examination of the strength ratios at the moment showed that the German and Italian forces were superior to the hostile forces at the front. Now was the moment to take preventive action, interrupt the assembly of the enemy forces through a counterattack, and delay his preparations for the continuation of the offensive. Rommel, therefore, decided to launch an attack with limited objectives while deciding on further action as the situation developed.

The attack was scheduled for the morning of 21 January. Various deceptive measures were taken to conceal the German intentions, including strict secrecy concerning the intended attack. Thus, regimental commanders were informed only one day before the attack was to start. Also, all vehicular traffic in the direction of the front was to cease during daylight from the fourth day preceding the attack. From then on, vehicular traffic toward the front was firmly restricted to the nights. These measures proved fully successful.

As part of the attack, the 90th Light Division, hitherto the Africa Division, was to break through the enemy lines on either side of the coastal road and to advance toward Agedabia. The motorized Italian corps was to follow immediately and was then to advance south of the Via Balbia, while the Africa Corps was to start out from an assembly area thirty kilometers south of the Via Balbia, in an enveloping pursuit designed to prevent the retreat of as many as possible of the enemy forces. The breakthrough by the 90th Light Division succeeded as planned, but Army Headquarters, to which the command had been upgraded on 21 January 1942, received no reports from the Africa Corps for a long time. The corps had run into a patch of deep sand and could only move forward with difficulty. The intended envelopment thus failed.

On 22 January, Agedabia was taken. In the following days, two attempts to pocket sizable enemy forces in the Antelat-Saunnu area failed as the German forces were too weak, and their intention had been recognized at an early stage. However, large quantities of materiel were captured in a surprise raid on Msus.

The Italian Supreme Command approved an advance as far as Agedabia—but not beyond that point. It feared reverses that might again endanger the Italian infantry divisions because of their lack of mobility. The employment of these divisions forward of the Marada-Marsa el Brega line was therefore not permitted. Rommel, nevertheless, persisted in his intention to take advantage of the opportunity of the moment. He advanced through the desert at the head of a specially organized battle group, and on the evening of 29 January, he captured the Benina quarter of the city of Benghasi. There, he received Mussolini's belated approval of his advance. On 30 January, Benghasi was captured and a brigade taken prisoner. In the following days, the pursuit was continued straight through the Cyrenaica. Derna was taken on 4 February.

The condition of the troops and the lack of fuel prohibited any attempt to attack Tobruk, so from 7 February on, the units were compelled to organize themselves in defense positions, with the north flank based on the Bay of Bomba in front of the British positions at Ain el Gazala.

Since it was now to be expected that the exhaustion of the troops on both sides would lead to a period of comparative quiet, Rommel flew to Rome and Germany to learn the intentions of the Italian Supreme Command and Wehrmacht High Command with regard to the conduct of the war in the Mediterranean in 1943. He found that practically no plans existed and that the Italians were even very averse to any offensive operations before autumn.

In April, Rommel therefore again took the initiative on his own responsibility. His opinion was that it was necessary to take preemptive action against a new offensive by the enemy, which he expected in June, with probably even stronger enemy forces than before. It was vitally important to the Germans to capture both Tobruk and Malta, the latter of which, as a naval and air base, interfered to an intolerable extent with German seaborne supply traffic. Since the German Air Force could support only one of these operations at a time, however, it would be necessary for them to take place in succession. Rommel considered it desirable to attack Malta first and then Tobruk. However, if the preparations for the capture of Malta required too much time, he thought it best to attack Tobruk first so that after that town had been taken and the border line from Sidi Omar-Bardia reached, all air force strength could be concentrated against Malta.* Rommel's suggestion was that the attack on Tobruk should open in the second half of May.

After some argumentation, this suggestion was approved, and it was decided that Tobruk was to be attacked first, because it would take longer to prepare for the attack on Malta. The supply situation was exceptionally favorable in May so that adequate quantities of fuel were available in Africa by the intended date of the attack, 26 May. To a considerable extent, the ammunition situation was also relieved.

Important lessons were learned in this phase of the campaign:

(1) In pursuit actions, success depends not so much on the strength of the pursuing force as on speedy action and, thus, to a considerable degree, on the personality of the commander involved. Relatively small units under young and energetic commanders (colonels) proved most effective.

(2) It is highly important to assign air liaison staffs to the pursuit forces. These staffs must be equipped with radios, so they can direct the close-support air units to worthwhile targets and, above all, so they can constantly report the lines reached to units in the air and thus prevent the air forces from bombing their own forces on the ground.

(3) Enveloping enemy forces is more difficult in the desert than elsewhere, since natural obstacles, such as rivers and so forth, where enemy manpower can be concealed, do not exist.

(4) It is not to be expected that any attempt to take the enemy by surprise through the use of deceptive measures that have once proved successful, such as air attacks on an enemy's headquarters, will meet with subsequent success.

(5) Terrain reconnaissance cannot be carried out too carefully.

---

*An advance into the interior of Egypt was thus not discussed.

f. *Late May—July 1942: The Battle of Tobruk and the Pursuits to El Alamein*

The operational plan underlying Rommel's new offensive was as follows:

(1) Frontal attacks by the Italian X and XXI Infantry Corps, which had been consolidated temporarily to form *Armee Abteilung Cruewell*,* to commence on the afternoon of 26 May in order to tie down the enemy forces in the Gazala position.

(2) Advance of the five mobile units under the personal command of Rommel in a move around the right flank of the enemy at Bir el Hacheim in order to wheel in on the rear of the enemy on the 27th and complete the envelopment by 28 May. Counting from the right, the five units were disposed as follows: 90th Light Division; German Africa Corps, with the 15th and 21st Panzer Divisions; the Italian motorized corps with the *Ariete* Armored and the *Trieste* Motorized Divisions.

(3) An attack on Tobruk, after the elimination of the bulk of the British Eighth Army's forces in the field.

These plans miscarried for a number of reasons. First, the two Italian infantry corps were too weak to tie down the strong enemy forces effectively. Initially, the enemy was admittedly taken by surprise by the forces that bypassed his southern flank. Then, however, the attacking column spread out fanwise as the result of the 90th Light Division turning northeast and the German Africa Corps turning north. At the same time, the Italian motorized corps, pivoting on the inner flank, was forced to move toward Bir el Hacheim and also reduce the speed of its advance. This fanlike disposition of the attacking forces greatly facilitated the defense.

On the evening of 27 May, the attacking mobile force, which had split into three groups, was in a critical situation and in serious danger of being encircled. Furthermore, up to 29 May, Rommel to a great extent was unable to exercise his command, having become separated from most of his radio stations. Supplies had to be routed around Hacheim and, as convoy forces were lacking, large amounts of materiel and numerous vehicles were lost.

Nevertheless, in spite of this unfavorable development, Rommel steadily persisted in his intention to take Tobruk. He

---

*A temporary organization commanded by a corps commander with a corps-type staff.

concentrated his forces once again, established a defensive front facing east, and from 1—6 June succeeded, one after the other, in eliminating a number of enemy strongpoints south and west of the main enemy position. In this way, and in coordinated action with *Armee Abteilung Cruewell*, he succeeded by 31 June in opening up a direct supply route. The route, however, was under fire during the daytime in most parts.

Having thus eased the situation behind the center of the enemy front, Rommel proceeded to eliminate Bir el Hacheim, a bastion in his rear. This point was well fortified with field-type positions and tenaciously defended; it was not taken until 12 June. Now the German Africa Corps advanced northward on Acroma, where it destroyed considerable armored forces by 14 June and threatened to cut off the two divisions in position in the northern sector. One of these divisions fought its way out eastward, while the other cut its way through the Italian forces by way of Bir el Hacheim toward the south. Now, at last, the road was open to Tobruk, which the British were determined to defend. A new deceptive ruse of Rommel's now proved successful. In the afternoon of 19 June, he moved his German Africa Corps eastward past Tobruk on the south, moved it back during the dark, and on the morning of 20 June, attacked the fortress from the southeast. On the following day, the fortress, with its garrison of 25,000 men and enormous stocks of supplies, was compelled to capitulate. On 23 June, Rommel crossed the border with the bulk of his forces, the 90th Light Division already having advanced to Sidi Barani.

Thus, the operational objective had been gained, and the time had arrived to release the bulk of the air forces for operations against Malta. The Italian Supreme Command and Field Marshal Kesselring, commander of the German Second Air Force, still intended to direct their attention to Malta, but Rommel believed that he now had an opportunity that would never recur of pushing ahead to the Nile. He was supported in his opinion by the German High Command and succeeded in getting his way. The attack on Malta was postponed, and the main mission of the air forces was to continue supporting the pursuit in the direction of the Nile.

On 28 June, Mersa Matruh was captured, and on 30 June, Rommel arrived with his thoroughly exhausted troops and only fifty serviceable tanks before the Alamein position, which was better fortified than any position he had hitherto encountered. Two attempts to break through this newly established British front failed on 1 and 10 July, whereas serious crises resulted from numerous counterattacks by the British between 15 July and the end of the month, the British directing their attacks chiefly against sectors of the front that were held by Italian troops.

Supply traffic again diminished considerably so that for this reason alone, if for no other, any new offensive was out of the question. It was found that Tobruk, as a naval base, had far smaller off-loading capacities than had been expected.

To hold the front of about seventy kilometers, new German units had to be transferred to Africa, and the 164th Light Africa Division, the Parachute Instruction Brigade, and the Italian *Folgore* Parachute Division were brought across by air and sent into action. Transport of the Italian infantry divisions from Libya took a great deal of time.

The more important lessons to be learned from this phase of the campaign were the following:

(1) Once again, several tactical surprise actions had been successful because methods were changed each time. On the other hand, the "dust deception" ruse was no longer effective. Rommel had had airplane propellers installed in a number of vehicles for the purpose of creating clouds of dust. These vehicles had been organized into a dust-producing platoon from which he expected good deceptive results. These, however, did not materialize.

(2) It is dangerous for a force to leave a major strongpoint in its rear unguarded, even temporarily. If the forces available are inadequate to envelop the strongpoint, strong reconnaissance forces capable of combat should at least be left to keep it under observation and, if possible, to contain it.

(3) The British minefields, the extent and distribution of which were unknown to the German command, and the mined zones in the Ain el Gazala position frequently compelled the command to make tactically disadvantageous changes in its plans. Minefields also proved a good substitute for terrain obstacles, of which there is a lack in deserts.

(4) Attack columns must be held together tightly, and units should only be detached for some separate purpose in cases of extreme urgency.

(5) Commanders at higher levels should not change their positions too frequently, even if the attack is progressing favorably. The commander definitely must designate some specific spot as his command post and must maintain that post as a fixed point, even if the situation is unclear.

(6) In the case of air attacks on enveloped strongpoints or bases, it is necessary to designate the targets to be bombed with minute precision in order not to endanger the attacking ground forces. This is particularly difficult under desert conditions.

(7) Supply columns are defenseless and require protection in convoys when the situation is unclear or confused; otherwise, they are apt to fall prey to enemy reconnaissance.

(8) In defense positions, tanks also should be dug in at once. This should be done in such a manner that they can drive out of the positions immediately if necessary. The space between the tanks and the surrounding ground in the trenches provides good protection for personnel against enemy fire and bombs.

g. *August—Early November 1942: The Battles Around Alamein*

At the beginning of August, the strengths on both sides were about equal. Neither the British Eighth Army nor the German forces had any appreciable measure of superiority. Nonetheless, it was clear to Rommel that time was working against him and that as soon as the enemy had brought forward sufficient reinforcements, he would launch a powerful counteroffensive. Rommel, therefore, did everything possible to improve the German positions, with particular stress on the use of mines, including air bombs that were buried and prepared for electrical detonation. He even had what he called "mine gardens" laid in the outpost area and had all battalion command posts surrounded by minefields. In distributing the forces in the northern half of the defense line, which he considered the most endangered and which were in the zone of the Italian XXI Corps, he placed Italian battalions and battalions of the 164th Light African Division alternately.

As soon as the supply situation permitted, Rommel intended to make another effort from the other end of the line to break through to Alexandria. However, for the moment, this was not possible, especially because of the fuel situation. Toward the end of the month, sufficient supplies of fuel would at last be available, if a large tanker that had left Europe managed to reach Tobruk. It was on this hope that Rommel—on the night of 30—31 August, supported by the Italian Supreme Command and by Kesselring—based his plan to break through the southern part of the front (which was held by weaker forces than the rest of the British line) and advance by way of Alam el Halfa to Alexandria.

The breakthrough—which was to take place at night, at two points, as part of two waves—was delayed until after daybreak, with the German Africa Corps on the right and the 90th Light African Division on the left, followed by the Italian motorized corps as the second wave. After the commencement of the attack, the commander of the 21st Panzer Division was killed, and the commanders of the German Africa Corps and 90th Division were both wounded. When this blow was followed in the morning by the news that the German tanker, on its arrival in Tobruk, had been torpedoed and sunk, Rommel intended to break off the operation. However, the chief of staff of the German Africa Corps induced him to continue the attack, which was making good headway. Field Marshal Kesselring also emphatically favored continuation of the attack. The attack soon began to move forward, but already by the evening of 31 August, the shortage of gas began to make itself seriously felt. Furthermore, a sandstorm, which had been blowing continuously, stopped after several hours, and enemy air attacks commenced with an intensity that had not been experienced before. At this stage, the five German and Italian divisions were behind the enemy front, where they were unable to move for several days. Meanwhile, they were attacked by the enemy air forces daily between 0700—1700 and 2200—0500 and suffered very heavy losses in personnel and materiel. Kesselring had promised to deliver 400 tons of fuel per day by air if necessary, but only a fraction of this quantity reached the troops. The reason for this was that most of the fuel was consumed by the transporting planes themselves on the long trip. It was only on 3 September that sufficient fuel was available for the Germans to commence moving back to the jump-off positions, which were reached on 6 September.

The following weeks were utilized mainly to further improve the defense positions. The three German and three Italian mobile divisions (the Italian *Littoria* Armored Division had arrived meanwhile) were organized in three tactical reserve groups, one German and one Italian division to each group, and held ready for action. Toward the end of September, heavy air attacks were launched by the enemy against German airfields and ground installations of the German Air Force, signaling an impending British offensive.

The offensive commenced at 2300 during the night of 23 October, which was a dark, moonless night.

In this offensive, the British Eighth Army employed the following:

    3 armored divisions
    7 motorized infantry divisions
    7 tank regiments, which operated independently

For the defense, Rommel had available:

    German forces— 2 panzer divisions
                           2 light divisions
                           1 parachute division
    Italian forces— 2 tank divisions
                           1 motorized division
                           1 parachute brigade
                           4 infantry divisions

The British had 1,200 tanks, among which were some of the latest Grant models. Rommel had 200 German and 250 Italian tanks, the latter of little value in combat. In the air, the Allied superiority was more pronounced than ever before, reaching a ratio of 10:1 at times in heavier-type bombing aircraft. In artillery and ammunition supplies, the enemy likewise was overwhelmingly superior.

The British attack opened on the northern part of the front, with the point of main effort shifting southward later, and by 29 October, the defenders had been forced to throw their last tactical reserves into the battle. The first attack was on the Italian strongpoints, and after taking these, the British enveloped the points still held by German forces.

The "mine gardens" referred to previously did not have the desired effect because many of the mines had been detonated by the artillery fire or during bombing attacks.

Although every inch of ground was hotly contested, a few kilometers being lost each day at the utmost, it was impossible for the Germans to hold the field permanently. Rommel, therefore, found himself forced to withdraw, if he did not want to risk complete destruction at Alamein. Consequently, he commenced withdrawing at the last possible moment on 3 November, contrary to Hitler's express orders. By that time, the enemy had broken through the German lines on a front of twenty kilometers. The 90th Light Africa Division had been moved to the rear previously to take up support positions at Fuka, where no defense line had been prepared owing to the lack of forces. The bulk of the Italian forces were captured, because no vehicles were available to render them mobile, as had been done with the Parachute

Instruction Division and the 164th Division. A great part of the German divisions succeeded in escaping capture.

From this phase of the campaign, the following important lessons were learned:

(1) The decision to attack on 30 and 31 August with a very insecure supply situation and the hope that the 5,000 tons of fuel would arrive was risky, but to persist in this decision, after it was learned that the fuel tanker was sunk, resulted in dire consequences.

(2) Once again, it was proved that only fully motorized units can be used in the desert.

(3) The defense would have been more successful if some of the mines laid within the main battle zone had been used in the rear to compel the enemy to change the direction of his drive frequently. In this way, the effectiveness of the main defensive weapon would not have been spent so soon.

(4) In transporting fuel by air, due allowances must be made for the fuel that the transporting planes themselves will consume.

h. *November 1942—January 1943: The German Retreat to the Border Between Libya and Tunisia*

The retreating troops, consisting almost exclusively now of German forces (particularly, the German Africa Corps and the 90th Light African Division), did not succeed in establishing a new line of resistance at Fuka, and even Mersa Matruh had to be abandoned on 8 November because of the danger of its being bypassed. Whereas the German forces had been under constant attack from the air by day and by night, these attacks gradually decreased temporarily because of the effects of heavy rains on the British airfields in the Nile Delta. A halt of one day was called at Sidi Barani, where considerable elements of the Parachute Instruction Brigade rejoined the army. They had set out to march through the desert on foot but had captured vehicles in a successful raid on British supply columns so that they were again mobile.

Thanks to the precautionary measures that had been taken to build up an effective antiaircraft defense there, the Halfaya Pass, which would have presented difficulties owing to the enemy superiority in the air, was crossed without serious losses.

The idea of defending Tobruk was weighed but rejected almost immediately, as it would have amounted to voluntarily accepting a siege. The retrograde movement continued, the Cyrenaica being abandoned up to the Marada-Marsa el Brega line, which was reached by the first combat units on 18 November. Here, at last, there was an opportunity to reorganize the units. Rommel expected a long stay at this point, since the enemy required time to close up his units from the rear and to move his supply bases forward.

On 28 November, Rommel flew to Hitler's headquarters, where he unsuccessfully suggested that the African theater of operations be abandoned. After his return from this trip, he decided to construct a rear position at Buerat. On 8 December, work on this position commenced. German units that required a period of rehabilitation, namely, the 164th Light African Division and the Parachute Instruction Brigade, were employed for this purpose as well as rear elements of the German Africa Corps and native labor, all under the direction of the commander of the 164th Light African Division. One of the main features planned by Rommel in this line was an antitank ditch in front of his positions, but owing to the lack of time and the inadequate labor forces available, he completed only parts of this ditch.

On 10 December, Rommel found himself compelled to abandon the Marada-Marsa el Brega line, since he feared that it would be bypassed. For the same reason, he abandoned the Buerat positions on 18 January 1943. At no point did he have sufficient armored reserves with an adequate supply of fuel to counter any attempts the enemy might make to outflank him.

Altogether, fuel supplies had become the major problem of this retreat. As no ships at all arrived in African ports, with the exception of a few military transports with a gross tonnage of 400 tons, the army was entirely dependent on air transportation for fuel supplies. On one single day, 200 tons were delivered in this way, but on all other days, the performances were far lower, rarely being more than 80 tons, and on one day, only 2 tons arrived. At any rate, the promised performance of 300 to 400 tons daily to be delivered by air was never achieved because of weather conditions and enemy activity. The fuel shortage was so serious that it was not even possible to take advantage of the favorable opportunities that frequently presented themselves to damage the pursuing enemy forces, since every drop of fuel had to be hoarded. Things got so bad that, in order to conserve fuel supplies, one motor vehicle was used to tow several others. This could usually be done along the coastal road, which was fairly level in most parts.

On 23 January, Tripoli was abandoned, and the retreat continued towards the Tunisian border, which was reached by the end of the month. Important experience gained in this phase of the African campaign include the following points:

(1) Before the battle began, German Army Headquarters should have combed out the transportation columns of the German Africa Corps and the 90th Division rigorously. Admittedly, these services were extremely limited, but a number of vehicles could have been obtained in this way to form a transportation reserve separate from the supply transportation services, and this reserve would have been available for the transportation of the infantry.

(2) In the face of enemy superiority in the air, it is impossible to maintain supplies for large units by air.

(3) During a retreat, all dispensable elements must be moved to the rear ahead of time, under a central command, and be directed firmly from point to point. This general rule equally applies in desert warfare.

(4) It proved even more difficult than in the previous year to intercept stragglers, since there were no natural features in the terrain that facilitated the establishment of straggler-intercept lines.

(5) Even during the retreat, the lack of engineer forces was seriously felt.

(6) The mining of airfields by the indiscriminate scattering of mines proved to be effective for a shorter time than had been expected.

(7) The Allied air forces made the mistake of attacking at regular times of the day, commencing their sorties at about 0800 each day, ceasing them about 1200, and then continuing them from 1400 to 1700. The German troops were able to take advantage of the intervals to increase the speed of their march.

i. *November 1942—March 1943: The Occupation of Tunisia and the Battles Fought in Tunisia*

Extraordinarily heavy convoy traffic in the direction of the Straits of Gibraltar was observed on 6 November 1942. The Italian Supreme Command in Rome—as well as Field Marshal Kesselring, Commander in Chief, South,* and commander of the German Second Air Force—immediately feared that the Allies

---

*Kesselring was given this title as the representative and coordinating head of German forces employed in the Mediterranean theater of war. Up to this point, he had command authority only over the air forces and, in a restricted degree, over the small naval combat units employed in the Mediterranean.

were going to land in force in French North Africa, a view which the Wehrmacht High Command, however, did not share. In fact, Göring, as commander in chief of the Luftwaffe, forbade the employment of strong air forces against the ship movements, which Kesselring had intended.

When the Allied forces landed in Morocco and Algeria on 8 November, they were met by French troops alone. The German and Italian control organs of the Armistice Commission were only a few hundred strong and were valueless for combat purposes.

It was only on 10 November that the Wehrmacht High Command ordered the Commander in Chief, South (who now, for the first time, was to take a part in ground operations) to occupy Tunisia. The only forces in Italy at that time were individual units of Rommel's army that were awaiting transportation to Africa. No integrated body of troops was available. For this reason, the occupation of Tunisia bore the imprint of an improvisational measure from the outset. The first unit to be moved there was a guard battalion of the Luftwaffe, which was transported by air. This unit was followed by a fighter group and Italian elements.

On 15 November, Lieutenant General Nehring,* as commander of Tunisia, assumed command over all units employed in Tunisia. He succeeded in gradually extending the occupation, which at first had been confined to the immediate surroundings of Tunis. By the end of November, Sfax and Gabes, in the south, were also occupied.

After the transfer of additional German units, the advance units of the British First Army, operating in the north, were pushed back to Tabarka and Medjez el Bab. The port of Bizerta was surrendered by the French without any resistance.

On 8 December 1942, General von Arnim, as commander of the newly created Fifth Panzer Army, assumed command over all army forces in Tunisia. Owing to the beginning of the rainy period, the Allied forces in Morocco and Algeria were unable to move their units eastward so that it was possible to consolidate the position in Tunisia by the end of the year and to establish an admittedly thin line of resistance against the British in the north, French forces in the center, and a U.S. corps in the

---

*Commander of the German Africa Corps from February to September 1942, initially as deputy and then as successor to General Cruewell.

south—in a general line east of Tabarka-Medjez el Bab-Fonduk-Faid-Maknassy. In the battles that took place on this front in early 1943, particularly in the central sector, a number of local German successes were gained.

Early in February, Rommel arrived with his army along the Mareth Line. This was a line of French fortifications at the former border that had been stripped at the demand of Italy after the defeat of France in 1940. The Mareth Line offered several advantages: it was only thirty-five kilometers long, it was protected by a continuous line of antitank obstacles, and its south flank was securely anchored on the almost impassable Matmata hills. Thus, it would only be possible to dislodge the German-Italian Panzer Army by an enveloping movement entailing a wide detour. On the other hand, there was the disadvantage that the supply route to Tunis, which was more than 400 kilometers long, could easily be cut by advancing U.S. forces, since the 100-kilometer section between Maknassy and Schott el Djerid was not protected owing to lack of forces.

To remove this threat to his rear, Rommel attacked at Faid on 14 February 1943, employing the bulk of his forces in the attack, some of them advancing by way of Gafsa, while the Allied forces were contained by elements at Fonduk. In this attack, he had available units of the Africa Corps—including the 21st Panzer Division and the 10th Panzer Division, which had been made available to him temporarily by the Fifth Panzer Army. Rommel succeeded in breaking through at the Kasserine Pass at Faid and in advancing to Tebessa, which he held for ten days. The American troops, which were still unaccustomed to combat, suffered considerable losses, and for the time being, the threat to the rear communications was removed. However, Rommel no longer had sufficient forces for the drive on El Kef that Hitler and Göring desired.

Since the German-Italian Armored Army under Rommel and the Fifth Panzer Army under von Arnim had now come into immediate tactical contact, a reorganization of the chain of command became urgently necessary. On 1 March 1943, the Army Group Africa was created and Rommel appointed as its commander. This army group was assigned the Fifth Panzer Army and the Italian First Army, the latter under the command of General Messer. Hitherto, the Italian First Army had been a part of Rommel's army. The African Air Corps, which also had just been created, was to cooperate with the army group.

On 6 March, the Africa Corps launched what was to be its last attack in Africa. The plan of this attack was to strike the

British Eighth Army on the flank while it was preparing for its new offensive. The armored units carrying out the attack were to operate from the Mareth Line. However, the general commanding the forces in this attack had no experience in this theater of operations, and the attack was halted by the heavy antitank defense of the enemy and, as a result, of the clear superiority of the enemy in the air. Heavy losses were suffered. Rommel had seriously doubted the chances of success from the beginning but was unable to escape the necessity of gaining time.

At the express order of Hitler, Rommel left the African theater of operations on 9 March 1943, General von Arnim succeeding him as commander of the army group, while General von Vaerst assumed command over the Fifth Panzer Army.

On 16 March, the British Eighth Army commenced its preparatory attacks against the Mareth Line, following up with the main attack on the night of 19 March. After crossing the antitank ditch, which was accomplished with difficulty, the British, with strong air support, succeeded in expanding the penetration. Nevertheless, a German counterattack on 22 March succeeded in recovering the greater part of the ground that had been lost. Then, however, on 26 March, Montgomery, employing two divisions, succeeded in breaking through the flank position so that the Italian First Army had no choice but to withdraw to the Akarit wadi, where the line was thirty kilometers long and well protected on both flanks.

The following are the main points of the experience gained in this phase of the campaign:

(1) It took some time before the command and troops of newly arriving units in Africa accepted and adapted themselves to the conclusions that had to be drawn from the overwhelming Allied air superiority. One of these conclusions is that in the face of enemy air superiority, the employment of massed armored units is doomed to failure.

(2) Once again, the value of speedy action was proved by the seizure of Tunisia, which took place solely with improvised means.

(3) The hopes centered on the maintenance of the short supply route from Sicily to Tunis did not materialize. The chance no longer existed of forcing the enemy to dissipate his air reconnaissance and air combat forces by using a number of sea routes. The enemy was now able to concentrate his air attacks against the one existing sea and land route of supply. For this

reason, the supply situation, which was eased temporarily in early 1943, worsened steadily. From the end of March 1943 on, transportation by large ships, which was the only way in which requirements could have been met, ceased almost completely.

j. *April—May 1943: The Final Battles in Tunisia*

The attack by the British Eighth Army against the German-Italian positions at the Akarit wadi began at daybreak on 6 April. Although the defenders were taken by surprise, the attack failed to penetrate. However, the increasing number of Italians who deserted showed how their morale was declining. In the German units, the shortage of ammunition for the artillery and special weapons was becoming more and more serious.

An attack by the U.S. corps in the direction of Gafsa-Fonduk resulted in a breakthrough at Gafsa and, in view of the threat to the rear that now developed, made the evacuation of the Akarit wadi position unavoidable.

In the rear of the Akarit wadi position was an extensive section of hilly country that offered poor protection for the west flank so that it was now necessary to retire to the Pont du Fahs-Enfidaville line, about 180 kilometers to the rear, and to abandon the intervening terrain to the enemy—almost without a fight. On 13 April, the Italian First Army moved into this line, which was about fifty kilometers long, and in doing so, for the first time made contact with the Fifth Panzer Army, which was holding the 120 kilometers of front extending from the coast in the north to Pont du Fahs.

In May, event followed event in rapid sequence. On 3 May, the British First Army penetrated as far as Mateur, which necessitated withdrawal of the northern flank of the German-Italian front to the area immediately west of Bizerta.

In the decisive and final attack, the enemy moved in two divisions of the British Eighth Army and directed his point of main effort at the center of the sector held by the Fifth Panzer Army, which he penetrated with strong air support at 1530 on 5 May. Tunis was captured by the Allied forces on the same day, Bizerta on 7 May. The southern flank, which was held by the Italian First Army, was not under such heavy pressure and was still intact when the western sector collapsed.

On 12 May, all resistance by isolated groups ceased. Crowded together on the Cap Bon peninsula, more than 250,000 men, more than half of whom were German, were taken prisoner.

The Commander in Chief, South, had intended consolidating all staffs and withdrawing all specialists—such as gunners, radio specialists, tank crews, armorer artificers, and so forth—in order to fly them to Europe and thus prevent the capture of at least the more important personnel. He was supported in this intention by the Italian Supreme Command but was prevented from putting his plan into effect by the deputy chief of the Wehrmacht operations staff, who intervened in April.

The following points concerning this final phase of the campaign in Africa deserve mention:

(1) The smaller the area becomes that is available to the defending forces, the more concentrated will be the effects of the force attacking it with superior power, particularly in the present age of long-range weapons. A superior air force can almost completely paralyze all movements on the ground.

(2) The brief descriptions of the individual phases of the campaign given in sections a to k show that the time chosen for attack varied; Rommel, for instance, preferred moonlit nights, while Montgomery chose a dark night at Alamein but, on the whole, usually commenced his offensives at various times of the morning. For the side that lacks air superiority, moonlit nights are particularly advantageous for attack, provided the troops have been adequately trained.

(3) The side that is weaker in air strength must restrict all movements of troops and supplies to the night, as otherwise, raids by enemy fighter-bombers along the roads would cause heavy losses and might even bring all movements to a complete standstill for a considerable period of time. Movement at night requires careful planning and organization. Even for vehicles traveling alone, it is advisable during the daytime to avoid main roads.

So far as the German side is concerned, the reasons for the defeat in Africa are to be found in the poor balance between ground forces, air forces, and naval forces. This poor balance resulted in the temporarily inadequate support rendered to the ground forces in November—December 1941 and the permanently inadequate support from the autumn of 1942 on, which was insufficient in spite of the self-sacrificing efforts that were made.

The lack of balance also resulted in the constant lack of adequate air and naval protection for supply transportation from Europe to Africa, the volume of which was adequate only once, in April—May 1942. This inadequacy again resulted in the ground forces always being short of supplies and later in their being handicapped by an acute lack of supplies of all kinds, particularly fuel and ammunition.

In operations in an overseas theater of war, in particular, a well-considered balance between the three branches of the armed forces is of decisive importance. Attempts should never be made to offset deficiencies of strength in one branch by an increased use and augmentation of another—which, in the present case, was the army.

III. SPECIAL FACTORS

8. Dust

a. *Effect on Troops, Weapons, and Equipment*

Men in the desert are constantly exposed to the effects of dust. This bothers the fighting man all the more because he has to endure it in conjunction with heat and the lack of water. There is no universal remedy against dust in the desert. Dust is a betrayer that enables observers, both from the ground and the air, to perceive every movement for great distances, even by individual vehicles.

Every footstep on the surface of the desert throws up dust and sand. Moreover, the almost perpetual winds carry along dust with them, generally in the form of dust columns as high as a house, which form themselves into whirlwinds and dust devils. In the beginning, the German troops in the desert suffered considerably from dust and had to fight against mental depression. However, they quickly became accustomed to it, so their fighting power was not affected to any appreciable extent. The dust in Africa does not cause any injury to health, since it does not contain any angular or sharp-edged particles that might lead to lung diseases. The eye inflammations caused by dust did not have any serious consequences. Wearing dust goggles proved helpful to the men, especially when traveling in the large clouds of dust produced by moving columns of motor vehicles. Therefore, all soldiers in the desert were equipped with a pair of dust goggles.

The effects of dust on weapons and equipment, including motor vehicles, was considerable in the desert. Dust had the greatest effect on motor vehicles, because the dusty air that was sucked into the cylinders attacked the cylinders and pistons and caused these parts to wear out quickly. Special air filters reduced the wear but could not prevent it altogether. In general-purpose cars (Volkswagens), the air intake openings were installed in the interior of the cars to give the engines purer air. In tanks, the air was sucked out of the battle compartment. In spite of this, the average lifetime of a Volkswagen engine in the desert was only 12,000—14,000 kilometers in comparison with 50,000—70,000 kilometers in other theaters of war. In the desert, it was necessary to change tank engines after about 3,500 kilometers, while they would last for 7,000—8,000 kilometers in Europe. To be sure, this was due not only to the effects of dust but also, to a considerable degree, to the necessity of driving long distances cross-country in low gears. The other parts of motor vehicles (such as brakes, chassis, and all parts that could be penetrated by dust) also suffered considerably more wear and tear than under normal conditions. It is not possible to give any figures on this point. What is certain is that motor vehicles in the desert need substantially more lubrication than in other theaters of war. No special greases and lubricants were used.

The barrels of guns, as well as all unprotected moving parts, were especially affected by dust. The wear on barrels, therefore, was considerably higher than in the European theater of war. Machine guns, submachine guns, and other small arms were the weapons most endangered, because inasmuch as they were used on the surface of the ground, they were especially exposed to the effects of dust. It was necessary, therefore, to protect all the movable parts of guns and equipment—especially the breechblocks—by such expedients as wrapping them up when not in use, covering them with shelter halves, or by other means. The barrels of artillery pieces and rifles had to be provided with muzzle protectors whenever they were not being fired. In view of the effects of dust, special importance was attached to the care of weapons and equipment, as well as to cleaning them frequently. Improvised dust-proofing devices have been discussed in detail in the former German field manual *Combat in Deserts and Steppes* and, therefore, no further mention will be made of them.

   b. *Effect on Combat Operations*

The generation of dust made it practically impossible to conceal marching columns. Dust clouds could be seen even at great distances and enabled one to recognize the size of the columns and sometimes even the type of vehicles (wheeled or tracklaying). On the other hand, the effects of dust were also taken advantage of for purposes of camouflage and deception. Dust was often created artificially in the desert, chiefly for purposes of deception. Rommel was the first to recognize the possibilities of this method, and he employed it up to the summer of 1942. However, even he often fell a victim to enemy deception measures.

Concerning the importance and effects of dust, Field Marshal Rommel said:

> On 13 March 1941 I transferred my headquarters to Sirte so that I could be closer to the front. In order to save time I attempted to reach this area by airplane. In the area of

Tauroga a sandstorm came up. The pilot of the airplane turned around, although I tried to get him to fly on. The trip was then continued by car. We were now forced to admit that we had had really no idea of the tremendous force of such a sandstorm. Huge clouds of a reddish blue hue obscured our vision and the car crawled slowly along the coastal road. Often the wind was so strong that one could not drive at all. Sand dripped down the car windows like water. It was only with difficulty that we could breathe through a handkerchief held in front of the face and perspiration poured from our bodies in the unendurable heat. That was the *ghibli*. In the silence I made my apologies to the pilot of my airplane. One Luftwaffe officer actually crashed with his airplane in the sandstorm that day. On 4 April 1941 I got underway with my combat staff at 0300 in order to bring the artillery battalions into their positions before daybreak. In the complete darkness we did not find the columns. On the next morning we repeated our attempt and were finally able to locate the artillery. Among other things, we ran into the rear of a British outpost area without knowing it. Although we only had three vehicles, of which only one was fitted with a machine gun, we drove up to the enemy at high speed while raising a great deal of dust. This apparently made the Englishmen nervous and they evacuated their position in great haste, leaving weapons and materiel behind.

In the description of the most important battles in the desert (see chapter II, section 7), further references appear concerning the effects of dust on combat operations.

c. *Effect on Tactical Measures*

During the first attack on Tobruk, dust had the following effects, concerning which we quote the following passage from Field Marshal Rommel's diary:

> The "Brescia" and "Trento" Divisions were supposed to attack Tobruk from the west and to raise a great deal of dust in the process in order to deceive and pin down the enemy. During this time the main attack group was supposed to swing around south of Tobruk in a wide arc through the desert and attack from the southeast. The dust which was thrown up deceived the enemy so thoroughly that he guessed that the attack would come from the west and paid no attention to the enveloping movement. When the enveloping group had reached its jump-off positions, their columns were struck by heavy British artillery fire. However, the air was soon full of heat vibrations and gusts of sand blew into the faces of the enemy. Good visibility soon vanished completely.

On 11 April the encirclement of the fortress of Tobruk was completed. The "Brescia" Division opened the attack. A great deal of sand was blowing and the British artillery could therefore not be expected to direct any aimed fire.

At about 1300 several enemy tanks moved past Ras el Madauer toward our lines. Because of the tremendous amount of dust, which moreover was being blown toward our positions, it could not be seen whether they were followed by any additional tanks and whether it was really a major attack. Therefore, I immediately committed all the antitank guns which were available in this area. It actually was a major attack and we succeeded in knocking out several tanks and halting the enemy advance.

Around 1800 on 30 April a new attack was opened against Ras el Madauer. Numerous Stukas cooperated with us. Soon the hill was hidden in thick clouds of smoke and dust. The visibility of the enemy was reduced to zero. It was impossible for them to deliver any aimed fire. Our attack led to a complete victory.

During the advance of the German Africa Corps from the Alamein position into the British rear area, the effects of sand varied. After the Africa Corps had replenished its supply of motor fuel and ammunition on the morning of 1 September, it began to move about 1300. At first, the attack made good progress in the violent sandstorm, which blew into the faces of the enemy. Unfortunately, the Italian divisions were far off and were unable to take advantage of the camouflage provided by the dust clouds in their advance. The vehicles and tanks toiled laboriously through the deep sand drifts that covered the attack areas. A fitful sandstorm raged all day and prevented the British Air Force from attacking in strong formations. When the sandstorm abated during the evening, the spearheads of our attack were engaged in stubborn combat with a strongly fortified enemy defense position, and the attack came to a halt. (Incidentally, reference might also be made here to the statements given in chapter II, section 7.)

The intervals that the advancing units were ordered to keep from each other to avoid dust varied according to whether the dust cloud was being blown in the direction of the advance or to the side. Moreover, since these desert expanses were, in general, easily traversed by all kinds of motor vehicles, it was possible to drive with gaps between the separate vehicles, thus reducing the effect of dust on the driver and his visibility. In general, we used intervals of fifty meters, both in depth and width. During the night, this interval had to be shortened for the sake of visibility to maintain contact with the man in front.

The generation of dust through the recoil of the powder gases in artillery firing was of no special importance for the detection of artillery positions, because the combat zone was always enveloped in dust clouds anyhow. The discharges of guns of especially flat trajectory with a low-barrel elevation—antitank guns—could be observed and recognized with particular ease by the enemy because of their characteristic dust clouds. Naturally, they also prevented the gun crews from observing the effects of their own fire.

d. *Effect on Aircraft and Their Crews*

Sand and dust had no appreciable, immediate effect on airplanes and their engines, since sand filters were attached to the intake valves. Dust had no effect on the efficiency of the engines, but nevertheless one had to expect more rapid wear and tear on the engines, since very fine dust particles were not entirely kept out by air filters. Even the special precautions taken during refueling did not always provide 100 percent security.

Very heavy sandstorms made flights practically impossible because of the extremely poor visibility when taking off. However, such storms were comparatively rare. The ghibli brought sand out of the interior at heights of as much as 5,000 meters; dust was still easily visible 100 kilometers out to sea and indeed occasionally was even carried as far as the European continent. This greatly hampered horizontal visibility, especially facing the sun. Sometimes visibility was reduced to below ten meters. On the other hand, vertical visibility was only slightly impeded. Nonetheless, it was only in exceptional cases that direct observation and aerial photography furnished satisfactory results about target details. In all airfields that consisted merely of sand, it was difficult and sometimes dangerous for several airplanes to take off and land together. When there was no wind, the dust remained hanging over the ground for an endlessly long time so that in spite of extensive improvisations, formation takeoffs failed in their purpose. Landings had even more unfavorable effects, since machines with empty fuel tanks simply had to land in case they could not reach an alternate airport. When the wind was blowing, airplanes took off with a slight crosswind, so the dust raised by the takeoffs would be blown to one side and not disturb the pilots behind. Difficulties also arose in dropping bombs on point targets, since the dust thrown up by the first bomb made it impossible to sight the target accurately. Although the breathing, sight, and other functions of men in machines were hardly disturbed by sand, radio equipment was more sensitive. Many radio failures could be traced to damage from dust. The most widely different methods were adopted to reduce the ill effects of dust at the airfields:

(1) By selecting surfaces that were somewhat grassy or crusty, even if they possessed other disadvantages.

(2) By laying out abnormally large airfields or several airfields located close together.

(3) By reinforcing the surface of airfields with asphalt or mats. Nevertheless, dust and sand also had certain advantages for observers and scouts. They made it easier for scouts to detect every movement (by visible tracks), even on trails and airfields. However, inexperienced crews often overestimated the strength of the enemy. The tracks visible in the desert sand also enabled one to recognize where enemy troops had passed, as well as the strength and objective of movements.

Besides the sand filters attached to the intake valves, no protective devices were installed either in the engines or in the airplanes themselves. On the ground, it was possible to protect airplanes, engines, and machine parts against sand only to a limited extent by the use of awnings. Repairs were made in repair tents.

The German Army and the Luftwaffe, in general, protected themselves successfully against dust in the desert by the most widely different means and by taking particular care of weapons, equipment, and machinery. In general, weapons failed because of dust sooner than engines. These failures, however, were not of vital importance.

9. Terrain

a. *Influence on Tactical Measures*

From the general description of terrain given in chapter II, section 4, it is apparent that, with the exception of places with deep sand and rugged valleys, the desert in the combat zone of the German troops was, in general, passable for both wheeled and tracklaying vehicles.

The influence of terrain on tactical operations is just as decisive in the desert as in other theaters of war. It is more difficult, however, to take advantage of the peculiarities of the terrain for one's intentions, since due to the lack of forests, cultivated areas, villages, etc., it is seldom possible for troops to approach and assemble under cover. Nonetheless, even in the desert, there are widely different opportunities to take advantage of the terrain. For example, troop assemblies can be concealed in ravines and valleys from ground observation and—to a limited extent—even from air observation.

In both attack and defense, the important thing was always to have reconnoitered the terrain carefully in advance. In attack, importance was attached to choosing ground that could be easily traversed by motor vehicles and, especially, which offered a covered approach, at least in part, through the utilization of terrain contours. The fact that the desert surface was easily traversed by motor vehicles made it easy to advance in light formations with few casualties, as well as to make all kinds of enveloping movements. In actual practice, few limitations existed on freedom of movement. Thus, it was also easily possible to shift the direction of an attack. During an attack, the tank battle always occupied the foreground. Attempts were made to compensate for the lack of good observation posts by sending out forward observers. Difficulties arose for attackers if they were compelled to use the southern portion of the desert proper, which in places was covered with soft soil. In its attack of 30 August 1942 from the Alamein position, the Africa Corps had to contend with these difficulties. Above all, many motor vehicles became stuck in the passes that led from the ridge of hills to the depressions and thus offered welcome targets to the enemy air forces. The failure of this attack can be attributed, in part, to the unfavorable ground, together with the overwhelming air superiority of the enemy, the weakness of our own forces, and the lack of motor fuel.

In defense, terrain was preferred that offered an opportunity to prepare reverse-slope positions echeloned in depth. Moreover, the efforts of the troops to "crown the heights," which was dictated by the desire to see farther into the country lying ahead, had to be constantly combated. Units naturally desired to have terrain in front and on the flanks that was not easily traversed by motor vehicles, but this wish could seldom be fulfilled. The following positions offered the most favorable opportunities for defense in the North African desert:

(1) At the El Alamein position, it was necessary to defend a strip of open desert and steppes sixty kilometers long by field fortifications. There were no possibilities for envelopment movements by major formations, since the position was blocked off on both flanks. In the north, it was protected by the Mediterranean. In the south, the position had direct-flank protection in the form of the northern edge of the Qattara Depression (Senke), which has only three easily guarded passes, namely, the one directly west of the Alamein position at

Munquar Abu Dweis, then along the trail between Mersa Matruh and the Quara Oasis, and along the trail between the Quara Oasis and the Siwa Oasis. Of these passes, only the first was actually guarded by minefields and troops; the two others, however, were utilized by the small sabotage teams of the Long Range Desert Group to penetrate the German rear area.

Furthermore, the sandy soil of the Qattara Depression itself, which was filled with salt marshes, hampered movements by major units. Farther to the south, the great sandy desert served as a barrier to the hinterland. The only passage between the steep edge of the Qattara Depression and the sandy desert led through the Siwa Oasis, which was fortified as a strongpoint.

(2) Farther to the west, the Marsa el Brega position was the first to offer good opportunities for defense again. Here, the area of steppes and desert south of the coast contains many salt marshes and dunes so that only narrow zones have to be guarded by field fortifications. The open desert begins south of the El Faregh wadi and extends to the area north of the Marada Oasis. The attackers, therefore, are forced to make a wide detour.

(3) The Tarhuna-Homs position east and south of Tripoli is flanked by the Djebel Nefusa in Tripolitania and takes advantage of the mountainous terrain, which is not easily covered by motor vehicles.* Since the mountains descend steeply to the west but gently to the east, it can be more easily defended from attacks from the west.

(4) The two positions farthest west, which are the ones most favored by nature, lie in southern Tunisia in the area of Mareth and Gabes. The former takes advantage of the heights of the Matmata Mountains and is protected against extensive envelopment in the south by the Great Eastern Erg (region of sand dunes). There is open terrain there in the form of a 25-kilometer strip between the coast and the Djebel Matmata (eighty kilometers wide between the southern end of the Matmata Mountains and the great sandy desert). The Mareth position could be enveloped along this strip eighty kilometers wide, as first became evident during the fighting around the Mareth position.

The Akarit position situated north of Gabes is partially protected along its front and in its southwestern flank by salt

---

*The word "djebel," when used as part of a place name, indicates that the area mentioned is near a mountain (CSI editors).

marshes, which cannot be traversed by major units, and in the northeast by the sea. During the fighting here, the British broke into the position at the places that were not protected by salt marshes and forced the defenders to surrender.

Between the five positions named above, there were also three more defense lines that were used by either the Germans or the British during the hostilities. These positions were without any protecting obstacles and were only established as the result of the combat situation at the time, when the area behind them had to be held by the defender for tactical reasons. These were the following:

The Sollum position had no frontal obstacles. All strongpoints had to be dug into the ground. The northern flank was protected by the sea; the southern flank was open and could be easily enveloped. This position was chosen out of necessity, since the Germans intended to hold Tobruk and since this position was the key to the coastal highway and the important Halfaya Pass. Mobile units were organized behind the defense front to repel any enemy attempts at envelopment by mobile operations. During the British offensive in the winter of 1941, the front of this position was pinned down and enveloped in the south by strong British forces.

The Gazala position west of Tobruk was selected by the British as an outpost area for the fortress of Tobruk. It had no frontal obstacles, was protected in the north by the sea, and was open in the south. In May 1942, Rommel surrounded this position in a wide enveloping movement.

The Buerat position east of Tripoli had one weak frontal obstacle (Wadi Zem-Zem). The northern flank was protected by the sea; in the southern flank were several wadis that could be easily overcome by an attacker. Occupation of this position was ordered by the Wehrmacht High Command for the purpose of defending the eastern outpost area of Tripoli. It was enveloped by Montgomery during the British offensive in January 1943.

The three positions mentioned above were thus of slight value for the defense. The terrain situated between all these eight positions is unsuitable for a lasting defense, because everywhere it contains more or less extensive areas of open desert and steppes. A defense in these areas, therefore, can only be conducted along mobile lines.

The fact that for a distance of 3,800 kilometers, there are only five natural defense positions of any use shows the great superiority of the attacker in desert warfare.

b. *Influence on the Construction of Field Fortifications and the Use of Weapons*

The German troops constructed only field-type fortifications in the desert. In building them, an effort was made to keep the upper slope at ground level to prevent the enemy from recognizing them too soon. Special difficulties arose in constructing positions for antitank guns and heavy antiaircraft guns (high superstructures). These positions had to be emplaced on the reverse slope. In places where this was not possible, the expedient was adopted of keeping these weapons in readiness in some place in the rear and not bringing them up to the position until they were urgently needed.

No experience was gained in the construction of permanent fortifications. However, it should be pointed out that the Italians laid out the fortifications of Tobruk so cleverly that they met with Rommel's unqualified appreciation. The Italian emplacements, which were level with the ground, were later introduced into the German Army as "Tobruk positions" and used both in Italy and on the Western Front.

From a purely technical point of view, it is extremely difficult to prepare field fortifications in the North African steppes and desert. Wherever the ground in the steppes is stony, it is very hard, because there is a layer of so-called surface chalk on the surface. This layer is formed when the rain water absorbed during the winter rises to the surface again during the summer and evaporates. During this process, the dissolved matter, such as chalk, silicic acid, etc., is separated again and cements the top layers into a firm crust having a thickness of from fifty centimeters to two meters. Under this surface chalk layer, there is a so-called lixiviation stratum that is especially soft and therefore easier to work. In constructing field fortifications, it is first necessary to laboriously blast away the surface chalk layer. Work of this kind can only be done if sufficient time is available. If a temporary defense system is being established, one has to be content with erecting positions built out of such stones as may be laying around or else use steep slopes or ravines and fissures to get at the lixiviation stratum quickly.

The surface chalk layer is of maximum depth in the steppe area, which extends about thirty kilometers from the coast to the interior. In the desert proper, its thickness and firmness diminish, and it is finally replaced by a gravelly crust a few centimeters thick, which is of only slight importance for the construction of field fortifications.

Naturally, foxholes and shelters can be dug in the loose loam or clay of the depressions. However, because of the complete lack of timber, construction work presents difficulties, so it is necessary to limit one's self to bare essentials. The use of sandbags and sand-filled gasoline containers is of great importance.

Difficulties with underground water have to be expected in the salt marshes, in which there is saltwater to a depth of about one meter, even when the surface is dry.

It is impossible to lay minefields in rocky soil; in ground consisting of loam, clay, and sand, it is necessary to bear in mind that the mines will be exposed within a few weeks because of the action of the wind or else be clearly visible.

Experience has shown that the effects of artillery and machinegun fire are substantially more intense on rocky ground than on soft ground. Shells fired with percussion fuzes do not penetrate the ground and therefore can have an especially strong fragmentation effect. If solid projectiles are fired, the effect is increased by the frequent ricochets.

If it is at all possible to construct fortifications—this requires time and a great deal of materiel—they provide especially good protection. As an example, it might be mentioned that on some occasions, the Africa Corps used dried-up cisterns (indicated on maps by the word *Bir*), dating from Roman times, as command posts, ammunition dumps, and shelters for the troops. These cisterns had a small influx hole in the upper chalk layer, beneath which were large square caves of about 100 square meters and larger in area that extended through the lixiviation stratum. The roofs consisted merely of a layer of surface chalk one or two meters thick and were not supported for a length of thirty-five meters. Nevertheless, they held out against heavy artillery bombardments and air raids.

How difficult it is to capture well-constructed fortifications—if they are resolutely defended—became evident during the siege of Tobruk, from April to November 1941; in the engagements around the desert fortress of Bir el Hacheim, south of Tobruk, in June 1942; and on the El Alamein front, from July to October 1942.

The most important fortresses of the North African desert were Tobruk, Bardia, and Mersa Matruh—as well as the Alamein position, which was constructed like a fortress. The three former fortresses served to protect coastal harbors; the latter was a barrier erected at the gateway to Egypt. All desert fortresses were built in such a way that their works cannot be seen from the ground; that is, they were built level with the ground, had low-wire obstacles, communication trenches that were mostly of concrete, and strong antitank ditches. At the fortress of Tobruk, the outer ring consisted of two lines of strong positions that were not built like bunkers with embrasures but were completely sunk in the ground. In some places, the works in the outer line were surrounded by an antitank ditch. This antitank ditch was partly covered with light boards and a thin layer of sand and stones so that its outline could not be perceived even at close distances. The average length of one work was eighty meters. The work itself consisted of several shelters, well protected with concrete, that together could accommodate a crew of thirty to forty men. The different shelters were connected by a communication trench with combat positions for machine guns, antitank guns, and mortars at their points of intersection. Like the antitank ditch, the communication trench, which was about two and a half meters deep, was also covered over with boards and a thin layer of earth, which could be easily opened at any desired point. The works were surrounded with strong wire obstacles, and the individual positions were connected by barbed-wire obstacles. The second line, which was about 200—300 meters behind the first, was of similar design.

The desert terrain had a great influence on the selection and use of the various weapons. It was found that one cannot have too many tanks in the desert, for because of the almost unlimited possibilities for using and deploying tanks, they bear the brunt of desert warfare. An abundant supply of antitank guns is necessary, since in view of the almost endless distances, reconnaissance naturally assumes special importance.

All guns should have the longest possible range, since the enemy can be seen even at a great distance, and it is necessary to get him accurately within your sights before he has you covered. Since there is very little cover and only a few reverse-slope positions in the desert, it is advisable, for the most part, to use only weapons and vehicles (including tanks) with a low superstructure. With tanks, it is especially important to have one that is fast, maneuverable, and equipped with a long-range gun. Then, the question of whether the armor plate is of greater or lesser thickness is of no vital importance.

In the course of time, mines acquired tremendous importance in the desert. They were generally used for furnishing unobstructed terrain with artificial obstacles. All fortresses, strongpoints, and fortifications were protected by minefields. In the course of the fighting, the employment of mines in the desert developed into a real art on both sides.

c. *The Tactical Importance of the Recognition of Vehicle Tracks by Air Observation*

The tracks of motor vehicles in the desert can be easily recognized in aerial photographs. Together with other observations, tracks were constantly evaluated for the following purposes:

- To discover troop assemblies and concentrations.
- To ascertain enemy supply routes.
- To determine whether terrain in enemy territory was passable for motor vehicles.

It was, therefore, often possible to detect enemy movements in the flanks and rear of friendly territory, especially the movements of the British long-range reconnaissance detachments that were operating in the area of the German-Italian front.

d. *The Use of Vehicle Tracks for Deception of the Enemy*

An attempt at this form of deception was made during the first few months of the African campaign, but it was later abandoned because the expenditure of motor fuel was disproportionate to the results achieved.

e. *The Use of Wheeled and Track Vehicles*

The only track vehicles used in the desert were tanks, guns mounted on tank chassis, and antitank guns. The armored personnel carriers and artillery prime movers were half-track vehicles.

Wheeled vehicles, the same types as were also used in Europe, were employed for all other purposes, especially to transport troops, equipment, and supply goods, as well as to tow guns in an emergency.

Whereas track and half-track vehicles were able to traverse all kinds of desert terrain, wheeled vehicles frequently had difficulties, especially in getting over sand dunes or steep slopes. It would be desirable to use only tracklaying or half-track vehicles in desert operations. Then, there would be no difficulties whatsoever in moving troops, weapons, equipment, and supplies—except in getting over salt marshes.

Instructions concerning driving in the desert are contained in the German field manual, *Combat in Desert and Steppes*. These instructions were written on the basis of the experience gained by German troops in the desert and can be described as very useful.

f. *Influence of Desert Terrain on the Development of New Tactical Principles for the Use of Motorized Units*

Since desert warfare is determined by the terrain and has to be carried out on a mobile basis, mobile engagements will be decided almost exclusively by motorized units. Combat in this open, unobstructed terrain must be carried on after the manner of a naval battle. Above all, it is essential to recognize the enemy's intentions quickly and react immediately to them. This is only possible if the chain of command is short. The commanders, therefore, must be stationed in the immediate vicinity of, or right among, the combat troops and should not be hampered in their decisions by orders from headquarters that are far from the front. In desert warfare, a unit commanded from a rear headquarters runs the risk of being encircled and annihilated.

To a large extent, Rommel's victories were based on the fact that he realized these tactical necessities of desert warfare and consistently acted accordingly, while the British adhered strictly to orders that they had received a long time previously and were no longer applicable to the existing situation. Rommel's successes diminished as he became more and more bound by orders from higher headquarters in Germany and Italy.

Other points must also be considered:

- Open country permits a rapid concentration of forces at the decisive point.

- Long-range weapons of all types are of decisive importance.

- The troops engaged in operations should carry along supplies for as long a time as possible, since it is often impossible to send supplies after them because their lines of communication are threatened. Supplies must often be sent to the troops by the convoy system.

Field Marshal Rommel expressed his own views concerning the influence of the desert terrain on the development of a new system of tactics for motorized units as follows: "The North African desert was probably the theater where war was waged in its most modern form. On both sides the brunt of the fighting was borne by completely motorized units, for use of which there were highly favorable opportunities in this level, unobstructed terrain."

Here, it was possible to really apply the basic principles for the conduct of tank warfare as they had been taught in theory before the war—and especially to amplify them. Here, out-and-out tank battles were fought between division-size armored units. Although the war slowed down into infantry and position warfare from time to time, its most important phases—the British winter offensive of 1941—42 and the German summer offensive of 1942—demonstrated the principles of full mobility. In desert warfare, against a motorized or armored opponent, nonmotorized troops can only hold their own in elaborately prepared positions. If such a position is breached or outflanked, a retreat means delivering up such troops to the enemy. The most troops can do is to resist in their positions to the last cartridge. During the withdrawal from Cyrenaica in the winter of 1941—42, practically all the Italian infantry and many German infantry units had to be moved out by shuttle traffic of a few truck columns or else march on foot. Only sacrifices by the motorized units made it possible to cover the retreat of the German and Italian infantry units. Moreover, Field Marshal Graziani's failures in the winter of 1941—42 were largely due to the fact that a large part of the nonmotorized Italian Army was helplessly exposed in the open desert to attacks by inferior numbers of completely motorized British troops. The weak Italian motorized forces could not engage the British with any prospect of success.

From the nature of warfare in the Libyan and Egyptian deserts, it is possible to derive principles that differ from those applicable in other theaters of war.

For instance, enemy units encircled in a pocket can be destroyed under the following circumstances:

(1) When they are nonmotorized enemy forces or forces that must give support to other troops that lack mobility.

(2) When they are enemy forces that are clumsily employed or forces whose commander has decided to sacrifice them in order to rescue other forces.

(3) When they are enemy forces whose strength is already broken so that signs of disorganization and panic are visible among them.

The following points are also worthy of mention:

(1) A commander should strive to concentrate his own forces with respect to time and space, while, at the same time, attempting to scatter his opponents and defeat them one at a time. The terrain offers unlimited possibilities for such action.

(2) Supply routes in the open desert are unusually vulnerable. A commander should protect his own supply routes with

all available forces and try to destroy or cut off those of the enemy. Operations across the enemy's lines of communications can often force him to break off an attack at another point, since the supply system is essential for combat and therefore must be protected above all else.

(3) The armored troops are the backbone of the motorized army. Everything depends on the tank; the other units are there merely to support it. Therefore, the battle of attrition against the enemy armored units must be fought as much as possible by one's own tank-destroyer units. One's armored troops should deliver the final thrust.

(4) Reconnaissance data should reach the command in the shortest possible time, and the command should make its decisions and convert them into action as quickly as possible. The commander who reacts the quickest wins the battle.

(5) It is the speed of one's own movements and the organizational unity of the troops that decides the battle and deserves special attention, since desert terrain places hardly any obstacles in the way of swift movements.

(6) It is of great importance to conceal one's intentions. Deception maneuvers should be encouraged by all means, in some cases to make the enemy commander uncertain and induce him to resort to caution and delay.

(7) When the enemy has been thoroughly beaten, the attempt can be made to exploit one's success by overtaking and destroying major elements of his defeated units. Here, too, speed in exploiting unobstructed desert terrain is everything. It is necessary to reorganize one's forces for the pursuit as rapidly as possible, as well as to reorganize the supply system quickly for the offensive units.

g. *Influence of Rainfall on Mobility in Desert Terrain*

In the steppes, heavy rainfalls occur only a few days in the year and in the desert proper even more rarely. A few hours after the rain has commenced, the wadis begin to fill with water. The water rises quickly and converts the otherwise dry valleys into raging torrents. In the first winter of the African campaign, the troops lost equipment and even men as the result of this natural phenomenon. In spite of the fact that the troops had been warned and were required to evacuate the wadis, there were still numerous living quarters in the valleys during the first big rainfall. The water began to rise during the night and washed away tents and motor vehicles.

The strong effects of the rain could be attributed above all to the fact that the troops were stationed in the mountainous area between Tobruk and Bardia, which was full of numerous steep slopes and deep wadis. In the area of El Alamein (winter 1942—43), the effects of the rains, which commenced at the beginning of October, were substantially slighter, because the terrain was more level and the troops had gained experience in the meantime.

After a rainfall, the water collects in depressions and lowlands and renders them impassable. In the steppes near the coast, large connected areas became impassable for traffic for this reason. It was even impossible to drive over the trails, since most of them followed the depressions. After one or two days, the water evaporated and the difficulty was removed.

In the desert proper, rain had less effect on the surface. The water collected quickly in clay pits and changed them into lakes that lasted for several days—in some places even for several weeks. However, the water only filled the deepest parts of the depressions, while the border areas, where the troops generally pitched camp, remained dry. On the whole, rain had only a slight effect on the passability of terrain in the desert proper.

Rainfalls, however, had considerable influence on military operations during the retreat from the Alamein position in 1942. Near Fuka, one panzer unit became stuck fast so that the tanks had to be blown up. However, the rain bothered the British even more. They could not carry out their pursuits because of the mud, while the German troops continued their retreat along the coastal highway without interruption.

10. Water

   a. *General*

The water supply for the German troops in Africa was never a troublesome problem; therefore, it did not influence or hamper operational decisions. The chief reason for this was that there were always enough wells available. Moreover, in the Africa Corps, the entire water supply for the German troops was centralized and was under the supervision of the corps surgeon.

   b. *Requirements for Troops and Vehicles, Economy Measures, etc.*

During active operations, four to five liters of water were provided as a minimum supply for one man per day. This quantity was only issued to the field kitchen, which used it for cooking and making tea and coffee. Pure water was never issued as drinking water. During quiet periods, unlimited amounts of water were issued to the troops whenever possible.

The amount of water required daily by a battalion with an actual strength of 600 men—not including water for motor vehicle radiators—was about 3,000 liters. The amount of water required by the medical units was fixed at twice as much per man per day, i.e., about eight or ten liters. The amount of water required for the radiators of motor vehicles varied according to the types of engines and fluctuated between three and ten liters per vehicle per day. The figures for the water rations were generally the same at all times of the year.

There were no important effects caused by evaporation. No appreciable losses occurred because of evaporation, since the water was poured directly into covered containers or tanks at the standpipes and remained in them until it was used. Two other points deserve mention:

- The problem of keeping water cool remained unsolved.
- There is no data available concerning the relation between water requirements and evaporation at different times of the year and in different temperatures.

In combat units serving in the desert, the water consumption was generally so limited that the men even had to go without washing almost entirely. Each man usually received a canteen (three-fourths of a liter) in the morning filled with coffee or tea, which had to last the whole day. Normally no water was used to clean vehicles and equipment.

c. *Water and Motor-Fuel Requirements*

In all combat operations, our chief concern was about motor fuel and not about the water supply. Only the garrison of Halfaya suffered severely from the lack of water after its well had been destroyed by gunfire.

d. *Tactical Importance of the Presence of Water Sources*

Even in the operations of advance detachments, the motor fuel supply was of more importance than the water supply. This is especially remarkable if one compares the required quantities. For example, a tank needs an average of 50 liters of motor fuel for 100 kilometers, whereas, the three crew members need only an average of 12 to 15 liters of water per day. Added to this is the radiator water, averaging about seven liters. In a daily run of 100 kilometers, therefore, each vehicle requires 50 liters of motor fuel and 22 liters of water.

No great lack of water developed because the fighting was always in the coastal area. In numerous places in the coastal dunes, there were sufficient supplies of fresh water to take care of the needs of the troops. Above a stratum of salty underground water, there was a more or less thick stratum of fresh water that was constantly increased during the winter by rainfall and which did not evaporate in summer as much as in areas farther from the coast. This water was collected by laying out an extensive network of drainage ditches and by careful pumping. If too much fresh water was removed, the salt water underlying it forced its way in and rendered the installation useless.

There are water sources of this kind in the sand dunes along the entire western coast of Egypt from Sollum to El Alamein. In addition, there are a number of wells in the solid rock in the coastal plains that yield fresh water.

East of Sollum and extending as far as Ain el Gazala, there is a strip of steep coastline that has no continuous ridges of sand dunes and therefore contains only a few water sources of this kind. Here, it was necessary to dig for water in the bottoms of the wadis that opened on the sea, but only small quantities could be found, and it was of poor quality. The water-supply situation of the German troops, therefore, was at its worst during the siege of Tobruk and during the defense of the Sollum front. During this time, the troops were given water that was barely palatable and that contained one gram of cooking salt per liter. Some times, it was only agreeable to the taste if used in hot coffee. The result was a great increase in the number of intestinal diseases.

Farther to the east, the water supply becomes better as one approaches Cyrenaica. Here, there are several large and efficient water sources that assure an ample supply for fairly large numbers of troops. In Cyrenaica itself, there are many abundant springs in the chalky formations of the soil that can meet any requirements with respect to water. The coastal area from Benghasi contains numerous deep wells and surface wells and is therefore adequately supplied with water.

West of Agedabia, the coast becomes flat again and is enclosed by broad ridges of sand dunes that contain an adequate number of water sources. Finally, in the surroundings of Tripoli, there is an extensive stratum of underground water, which is one reason why this region, relatively speaking, is so thickly settled. Here, water is pumped mostly by windmills.

All these water sources are limited to a narrow strip of the coast. Outside of this strip, there are no springs or wells whatsoever that could be of any use for supplying major troop units. There are merely cisterns, most of which no longer contain any water. Only a few modern cisterns produce water into the summer. From August to November, even these cisterns are generally empty.

Occasionally, there are water sources (indicated on the maps) in rubble-filled wadis situated below steep slopes. Usually, wells have also been dug there, but they are always dry. If new wells are dug, only damp sand will be found. These wells are dug by the men in camel caravans, who obtain the necessary water for their animals by this artifice. Water sources of this type are worthless for supplying a military unit.

Large amounts of water are found also in the chains of oases along the nineteenth parallel, which includes the Siwa, Giarabub, Gialo, and Marada Oases. The artesian wells found here have their source in an underground river that is fed from the rainwater falling in the Sudan and which flows northward underneath the desert. Since the above-mentioned oases are situated below sea level, it cannot be assumed that the underground river flows any farther to the north. It is also impossible, therefore, to obtain any fresh water by deep borings in the area north of the oases chain.

The above account shows that the only reliable place for obtaining water was along the strip of coast where it was naturally available. If military operations had been carried out in the interior of the desert, water would undoubtedly have been the decisive factor that would have determined the strength of the troops and their radius of action. As it was, however, it was always possible to make use of the water sources, which were most favorably situated with respect to the front, although it was sometimes necessary to transport the water as far as 100 kilometers. In view of the Germans' bad situation with respect to motor fuel, traveling this distance was certainly a considerable burden. For drinking purposes on the Alamein front, the new expedient was adopted of making use of a temporary lake, which was formed of rainwater from the southern part of the front and filtered by equipment.

Of the large oases, only Siwa and Giarabub acquired any major tactical importance. The Siwa Oasis enabled the Germans to close off the only passage between the steep side of the range and the sandy desert during the fighting on the Alamein front and thus protect their flank. The Giarabub Oasis was used by the British as a strongpoint and base for their attack against the encircling ring around the fortress of Tobruk in November 1942.

e. *Assignment of Engineer Troops for Water-Supply Services*

Engineers were only used for obtaining water in the beginning; the work was then assigned to water companies. The water-supply company of the German Africa Corps had the following equipment:

(1) Several centrifugal pumps, driven by mobile power generators, with a lifting height of at least ten meters.

(2) Several mobile water-purification units.

(3) A motor-driven water filter with a diaphragm for the purpose of obtaining sterile water (not to be confused with chemically pure water). All the above equipment was used in all the units of the German Army and therefore was not designed for Africa alone.

(4) After late summer 1942, water distillation plants for the purpose of obtaining drinking water from seawater were also added.

(5) A laboratory, the chemist of which had to pass on the wells and analyze the water in them.*

Water was divided into the following categories according to quality:

No. 1 Water, which could be used for all purposes.

No. 2 Water, which could be used for cooking.

No. 3 Water, which could be used for motor vehicles, washing, etc.

f. *Well-Drilling Equipment*

On the whole, no experience was gained concerning well drilling. The Abyssinian wells, with which the troops were equipped, did not prove suitable for this purpose.**

Some of the water-supply companies were equipped with heavy, mobile, deep-boring equipment, Salzgitter-type (Meissel percussion drills). Their size and efficiency are not known. They were not actually put to use and probably would not have been

---

*It should be borne in mind that too large a concentration of magnesium sulphate in the water may lead to intestinal irritations and thus intestinal diseases (author's note).

**These wells were perforated pipes, driven into the ground, that gathered ground water.

successful. If the fighting had shifted to the area south of the nineteenth parallel, the Germans would have been in the region of the static underground water stratum and probably could have used the heavy-boring equipment to advantage. It probably would have taken several weeks to make a boring of this kind.

The Benoto drill—a drilling scoop attached to a heavy plummeting weight, on the underside of which were sharp, movable steel claws—proved useful for operations in the coastal region. In operation, the drop weight fell freely onto the ground, whereupon the claws dug into the earth and then closed automatically. When the drop weight was raised, the earth held in the claws was brought up with it and then disposed of when the claws were opened. The drill was named after its French inventor and was manufactured in Germany on a royalty basis (both heavy and light models). With the heavy Benoto drill, it was possible to dig wells fifteen to twenty meters deep in one day in loamy or gravelly soil, for which it was especially well adapted. However, it was impossible to drill below the underground water level with it because the force of its fall was greatly broken by the water. The small Benoto drill was chiefly used for test borings. It drilled a hole about ten meters deep and seventy centimeters in diameter in one day. The principle according to which it was built was good, but unfortunately, the construction was imperfect so that there were often operating difficulties. In particular, the wire cables were too weak and were subjected to much wear and tear at the guide pulleys. Nor were the claws able to stand up under the heavy strain. If these faults were eliminated, this drill might prove a very useful tool in desert warfare.

The water distilling company was equipped with six water distillers mounted on trucks. This unit was employed only in special cases, since its fuel consumption was disproportionately high (one liter of fuel was required to obtain ten liters of water). During the German advance toward Egypt, this unit performed very important service, since all the wells had been contaminated with neat's-foot oil by the British during their retreat.

g. *Method of Distribution*

As a matter of principle, all water was issued in canisters from the mobile water-purification unit, which served as an intermediate container. At the end of the water purification unit was a hose fitted with twelve faucets, which made it possible to fill twelve canisters at the same time. As a matter of principle, all water—both drinking water and radiator water—was carried in canisters. Bakery companies were not issued any fresh water, since by baking with seawater, it was possible to save salt.

The standpipes were manned by teams from the water-supply companies. The water was distributed according to a table prepared by the quartermaster according to the personnel strength and vehicles of the various units. The water was taken away by the supply vehicles of the units.

h. *Pipelines*

If a position is occupied for a fairly long time, it is advisable to lay water pipes, since this saves considerable transport space. Aluminum pipes are recommended, since they can be laid quickly and connected with each other by a bayonet catch. Of course, strict supervision of the pipeline and standpipes is urgently necessary. When moving into the Alamein position, we made use of a British pipeline. In spite of strict orders, it was impossible to prevent it from being illicitly tapped. The available water supply was therefore soon exhausted.

11. Heat

a. *General*

Even the very high temperatures during the hot seasons of the year failed to impair the efficiency of the German troops to any noticeable extent. This is shown particularly well by Rommel's summer offensive from May to July 1942. The achievements of the troops refuted the former opinion of the Italians that European troops could not stand up under high temperatures. Even the British showed no regard for the heat; they attacked at Sollum in June 1941 and were planning a new offensive for June 1942. The most important thing for standing up successfully under hardships of this kind is the toughness of the troops and the good examples set by their leaders of all ranks, who as a matter of course should share all deprivations with their subordinates.

Experience has shown that men under thirty-five are best able to adapt themselves to a hot climate. Acclimatization is easier in the open country than in cities.

b. *Effect on Unaccustomed Troops*

Newly arrived units had low fighting power and many losses through sickness. The heat paralyzed the men's willpower and thus also their powers of resistance. The diseases themselves were probably more a direct result of the soldiers' diets (bad water, canned food) and hardships than of the heat. The sudden temperature drop in the evenings, causing abdominal colds, was also an important element.

c. *Effect on Tank Crews*

In tanks, the effects of heat—measuring 45 degrees on the gyrostatic thermometer (113 degrees Fahrenheit)—was naturally much greater on the men than among other troops. The heat became unendurable for men in combat, when the hatches had to be closed because of artillery fire and when the engines and ventilating systems had to be shut off during pauses in the action because of the lack of fuel. Nevertheless, the German tank crews held out under even these temperatures.

d. *Measures Taken to Avoid the Noonday Heat*

In combat, it was generally impossible to make any allowances for even the greatest heat. During quiet periods in position warfare, however, there was always a noonday rest period of at least three hours. Cold rations were issued to the troops at noon and hot rations at night.

e. *Special Equipment for Protection Against Temperature Variations*

Measures were taken against temperature variations by requiring warm clothing and bellybands to be worn after sundown. Each soldier had three blankets, one of which had to serve as a ground sheet for protection against the night dampness. For wounded men, four blankets were required (two for covering the men and two for placing underneath them).

f. *Types of Shelter*

In combat, the troops had no shelter whatsoever. They either spent the night in their position or else in or underneath their vehicles.

Wall tents for four men, with a double top awning, were used chiefly as shelter. They proved thoroughly useful and met all the requirements of desert warfare. At the beginning of the campaign, low one-man tents, likewise provided with an awning, were issued, but later there was not a sufficient number of them available. There were also larger tents for ten or twelve men. However, these could be used for the most part only in rear areas, since it was practically impossible to camouflage them. The Italian field hospital tents also proved highly successful.

g. *Comparison Between the Efficiency of Troops in the Tropics and in Temperate Climates*

After the men had become accustomed to desert warfare and after those who could not adapt themselves to it had been eliminated, the efficiency of the troops in combat operations was not inferior to that of troops serving in temperate climates. Even the drivers achieved normal efficiency, since the wind stream while driving cooled them off and refreshed them. In all kinds of physical labor, on the other hand, the heat greatly reduced efficiency. In this respect, the Italian troops were superior to the Germans. For this reason, they did most of the road construction work and mine laying. The detour road around Tobruk, for example, was built by Italian labor units. It constituted an outstanding achievement that German troops could not have accomplished in the same time.

h. *Effect on Materiel and Equipment*

The heat had no effect on guns, because they became so heated during firing that this more or less neutralized the outside temperature. No damage to war materiel and items of equipment was reported or was to be expected.

i. *Effect on Visibility*

In the interior of the desert about ten kilometers from the coast, the vibration of the air makes accurate observation practically impossible for a distance of more than one kilometer. All objects at a distance of one kilometer and more appear to move, and it is scarcely possible to decide whether a dark spot on the horizon is an approaching motor vehicle or a destroyed vehicle. At still greater distances, all contours become so blurred that one always thinks that one is surrounded by a sheet of water, from which certain elevations stand out as "islands." In the beginning of the war in the desert, during the first attack on Cyrenaica, several units fell victim to this deception. They tried to drive around what they thought was a "lake" and therefore did not reach their appointed objectives. On the other hand, the vibration of the air protected small detachments in the desert from being discovered by the enemy from the ground. Surveying work suffered to an extraordinary extent from the vibration. It was only in the morning hours (until around 1000) that satisfactory work could be done with aiming circles or theodolites.

j. *Effect on Airplanes in Taking Off and Landing*

For the Luftwaffe, great heat often had a considerable effect on takeoffs and landings. The density of the air was so reduced that the lengths of takeoffs and lengths of landing strips had to be increased by as much as 50 percent. Landing operations could be indirectly disturbed by strong air reflections in the heated ground layer, thus causing errors in the estimation of altitude.

If airplanes stand for a long time in strong sunlight, the polished metal parts in the cabin, especially the glassed-in cockpit, become so hot that they cannot be touched. Therefore, the cabins have to be covered with tarpaulins. For the same reason, strong sunlight entering the cabin during flight is also disagreeable for the pilots. Curtains relieved this difficulty but hampered visibility.

IV. MISCELLANEOUS
12. Cartographic Service
a. *General*

In Libya, the German Army was initially solely dependent on the hastily reprinted Italian maps (scale 1:200,000), since at first, good German maps were available only on the scales of 1:1,000,000 and 1:500,000 (which were mainly destined for use by the Luftwaffe).

The Italian maps were poor and often showed points of interest more than twenty kilometers from the coast as several kilometers off. In compensation, there were captured British maps that included Libya and Egypt. They were excellent and as a result much prized. Later, they were reprinted. The French maps, used for the first time in Tunisia, were serviceable but were later replaced by German maps. In 1941, the Luftwaffe received a special map on a scale of 1:400,000 and in conic projection, which fully met requirements.

b. *Reliability and Methods of Use*

Sufficient maps were available for the following regions:

Tripolitania (area around Tripoli): Italian map, 1:50,000
Cyrenaica: Italian map, 1:50,000
Egypt (western desert): British map, 1:100,000
Southern Tunisia: French map, 1:100,000 and 1:200,000

The lack of good maps made itself felt, especially in defensive operations. New photographs were accordingly taken of the area of the Marsa el Brega position and of the Buerat position in cooperation with the Luftwaffe, the cartographic detachment of the Africa observation battery, the cartographic section of the German Army High Command, and the Military Geologic Office. The following method was used. The survey battery drew up a net of artillery points (APs) by use of traverses that were emplaced at the triangulation points (TPs). The APs were clearly identified by means of ground panels. Then, the Luftwaffe took a series of aerial mosaics with 30 percent overlaps. The results were transferred from the aerial photographs to the AP diagrams by the Military Geologic Office. This work was more difficult than expected, especially since only a simple stereoscope was available for the purpose. Correct evaluation of the form of the terrain and of the condition of the ground on the basis of the aerial photographs required a great deal of experience. Errors occasioned by the angle at which the photograph was taken, the distortion at the edges of pictures, etc., had to be compensated for as well as possible. After rough drafts had been finished, clean copies were made by the cartographic section, and these were reproduced by the printing section. In this way, maps were produced in a short time that fully met the requirements of the forces. They gave an accurate picture of the tactical conditions of the terrain, permitted orientation, and contained correctly surveyed artillery points.

In respect to tactics, the road maps made by the Military Geologic Office constituted an important supplement. The Italian, British, and French maps were redrawn and reprinted by the Mapping and Surveying Service and then distributed to the forces by the cartographic centers in Africa. The maps were always available in adequate quantities, with the exception of temporary scarcities brought about by the difficulties of supply.

The maps and documentary material of the Military Geographic Branch were used by all headquarters down to the regiment. They were used for the initial basic orientation in the conduct of operations and as such were indispensable. In addition, they contained a great deal of important information, especially in their description of road conditions. However, since only such documentation as was available at the beginning of the war could be used, there were gaps that only reconnaissance could fill in. The colored hachures on maps, indicating the passability of terrain or at least the limitations on its passability, were important.

13. Camouflage

Camouflage is very difficult in the desert and, in many cases, impossible. During the day, it was impossible to camouflage the movements of troops and columns from air observation. In the neighborhood of the front, the troops could only with difficulty be camouflaged from ground observation. The unavoidable dust clouds they raised betrayed any movement.

Nevertheless, troop concentrations can be camouflaged, if great care is used. Depressions in the terrain will have to be exploited for this purpose. All vehicles will have to be covered with camouflage nets and vegetation (camel's thorn) attached to the nets. Another means of camouflage, although on a small scale, is to seek out shifting shadows. In the open desert, all vehicles will have to be dug in as deep as possible. It goes without saying that the vehicles will have to be placed at as great a distance from each other as possible.

Although in most cases it will not be possible to camouflage the presence of vehicles and weapons, skillful camouflage can conceal the type of these vehicles (the arm of the service). It would be altogether wrong to resign oneself to the viewpoint that camouflage in the desert is useless. What has been said above is also valid for the troops in position.

14. Evaluation of the Enemy Situation Through Aerial Photographs

The main emphasis in aerial photography—in addition to determining the traffic on the Suez Canal, in the ports, on the roads, important paths, in the villages, and on the airfields—was to provide details on the troops and positions of the enemy. But to check on our own camouflage, from time to time, the German positions were photographed from the air.

Little by little, it was possible to improve the methods of evaluating the results of the aerial photography. As the enemy's air superiority increased, however, the possibility of taking aerial photographs diminished more and more. Nevertheless, aerial photographs are indispensable. The evaluation was done by the Luftwaffe Photographic Section in the usual manner.

In taking aerial photographs, dark-yellow filters were necessary. Developing required the addition of hardening material, since otherwise, the coating of the film melted, and there was the danger that microbes would damage it. In addition, dust-proof vehicles were used for the developing of the film. Relief maps were also produced, but the need for them was not great, since the differences in altitude in the Libyan and Egyptian deserts is inconsiderable.

15. Visibility at Night

In the North African desert, the nights are much darker than in more northern latitudes. Thus, it was often difficult to find a tent in one's own camp at night because it was impossible to orient oneself. As a result, it was also impossible for the enemy to detect troop movements taking place at night, unless there was moonlight or parachute flares were dropped by the air forces. On nights when there was a moon, the visibility was approximately as good as that prevailing in Central Europe at the fall of dusk. During the daytime, on the other hand, it was possible to see very far in the desert.

16. Choice of Camp Sites

In the main, camp sites were selected for their good protection from air attacks. Preference was given to depressions in the steep sides of the djebels, the banks of wadis, and the dune districts. Only the latter, however, offered an opportunity for troops to dig in vehicles and tents and to camouflage them. (This subject is taken up in detail in chapter III, section 9, "Terrain.")

17. Selection of Battle Sites

In the attack, the German Army preferred easily passable, open desert for a battle site, because it made it possible for the armored forces to exploit their mobility and combat tactics, which were superior to those of the British.

In defense, the positions selected depended to a large extent on the terrain. Positions had to be difficult of access and provide flank protection; otherwise, on the whole, they were useless for defense. Regarding selection of positions, see chapter III, section 9, "Terrain."

18. Time of Day Selected for Combat

Field Marshal Rommel preferred to start fighting on moonlit nights or at dawn. The fighting often lasted until dark, unless it was broken off sooner for tactical reasons. There were no lulls in combat, even during the great heat prevailing around noon.

In general, the Germans carried out no night attacks. On the other hand, at the end of October 1942, the British carried out their large-scale attacks on the German positions at Alamein exclusively at night. They used parachute flares to illuminate the battlefield and tracer ammunition to show their troops the direction of the attack and sector boundaries. The British commander in chief, Montgomery, preferred night combat. The German and Italian troops used the night for marches into their assembly areas, for large-scale shifts of forces, and for surprise. These night regroupings could almost always be carried out unnoticed by the enemy and, especially during the winter battles of 1941—42, came as a big surprise to the British.

19. Influence of the Desert Climate on Daily Service Routine

The daily routine in the desert did not substantially differ from that in other theaters of war. When not in combat, the troops were, as far as possible, given a lengthy rest period during the great heat around noon. The maintenance troops (workshop companies) were also given a noon rest period. Supply traffic could not afford to take these rests.

20. Special Problems of the Technical Services

Through the fighting in the desert, the German technical services were faced with the following special problems:

a. They had to keep the motor vehicles in running order, which, because of the conditions in the desert and the long supply routes, were exposed to extraordinary wear and tear. For this purpose, an increased number of repair shops were available. It is probable, however, that the problem would not have been solved had the stock not been replenished by captured British materiel from time to time.

b. The water supply had to be safeguarded. The troops specially assigned to this task (see chapter III, section 10) did their work well.

c. A detour had to be built around the besieged fortress of Tobruk. This task was undertaken by Italian labor forces and carried out successfully.

d. The Capuzzo-Mersa-Matruh-Alamein railroad had to be operated.

The above statements may give the impression that the technical problems encountered in the desert are less difficult than those in other theaters of war. This is not the case, however. On the contrary, many of the tasks that were necessary were so difficult that they were abandoned in advance. Among others, these included the extension of the railroad beyond Capuzzo, and at least up to Tobruk, and the procurement of steam locomotives—as well as the construction of a fortified defense position with bunkers at the Alamein front.

21. Influence of Light, Shade, and Sandstorms on Combat

The intensive sunlight had a dazzling effect that was particularly disturbing when the light was reflected by bright dunes. It was therefore absolutely necessary to wear sunglasses. In the afternoon, it was impossible to tell vehicles from tanks at a distance of one to two kilometers and up. Favorable times of day were the early morning and the afternoon hours from 1600 to 2000.

In the desert, range estimation cannot be depended upon. In most cases, the distances will be underestimated. It was therefore advisable to use a range finder. During sandstorms, visibility was nil; moreover, visibility was influenced by sand driven by the strong winds (from five meters per second in intensity). From the air, dust produced an effect like local ground fog. During a sandstorm, fighting was impossible. In addition, there were isolated instances of mirages. In spite of these problems, however, adaptation to the visibility conditions prevailing in the desert was made relatively soon.

22. Influence of Darkness on Radio Communications

Frequently, radio communication with Europe was interrupted for hours at a time during the nights because of pronounced fading.

23. Wind

The wind, constantly blowing near the coast, was regarded as very welcome by the ground troops; it made it much easier to tolerate the heat.

For the Luftwaffe, the fact that the general direction of the wind was interrupted by periodic land and sea winds on the coast (during the day from sea to land, during the night from land to sea) was of importance. Because of the temperature differences between the water and the desert, the winds on the coast are stronger than on the European continent. Moreover, they increase with the altitude. During sandstorms, the wind blew up to 150 kilometers per hour above 3,000 meters (at the height of storms). On the ground, the gusts blew at the maximum rate of sixty to eighty kilometers per hour. Inland, the wind is weaker. Very often, the direction of the wind is constant, especially on the coast.

No data can be given on the wind velocity at various altitudes for comparison with velocities elsewhere, because no statistics are available on this.

24. Special Equipment and Procedures for Aircraft Crews

It was always advisable to try for an emergency landing, because the emergency equipment for desert landings, the armament, the radio, and the equipment for protection from the sun and cold nights was ample. Moreover, aircraft that made emergency landings were easier to locate from the air than those that crashed. Furthermore, the terrain was favorable everywhere for emergency landings, even for aircraft with stationary landing gears. Even a belly landing was preferable to bailing out. There is no case of crews having been lost in the desert; the "desert emergency squadron" played an important part in this. However, German aircraft were seldom employed in the interior of the desert.

The briefing of crews on the native population and its customs was kept to a minimum and was generally limited to the crews of single-motored aircraft (fighters, Stukas, close-reconnaissance planes). The crews of long-distance reconnaissance planes and multiengined bombers and the parachutists to be dropped behind the Tunisian front were briefed on the basis of need. It was possible to keep the briefing short, because the Arabs throughout had a friendly and helpful attitude. The sympathetic attitude of the Germans, which was soon known in the desert, was calculated to encourage friendliness. Small gifts and visits to the sheiks helped.

The crews were systematically briefed about the emergency measures to be taken to survive in the desert. Briefings were based on certain cases and experiences. On this occasion, the following questions were discussed:

> Where are their oases, who occupies them, and what is their capacity?
> Where are the other water holes?
> What is the attitude of the local sheik?
> Should one assume that he is in touch with the Allies?
> Where are the nearest enemy forces?

It was difficult to live off the land, since the natives had only just enough for themselves. As a last resort, the hunting gun included in the special equipment was available. Because of the scarcity of game, success in hunting depended largely on the skill of the crew. Water was obtained by spreading tarpaulins during the night to catch the morning dew. Rewards in gold were given to the Arabs for the rescue of German soldiers.

Distress signals consisted of signal ammunition and smoke signals, which had to be used at the right moment when the crew of a plane overhead was able to see them. Large aircraft carried an emergency radio transmitter. In addition, if necessary, fuel and oil were to be set afire. The crews had been told to remain near their planes, if possible. Only if they were close to an inhabited place were they to start marching immediately. The aircraft had special equipment for the desert. This consisted of the following items:

> 2 liters of bottled water
> Emergency rations (hard sausage, hard biscuits, cola, Dextro-Energen,* Pervitin, cognac)
> Gasoline stove
> One-man tent, sleeping bag, rubber mattress, linen cloth to protect the head from the sun

---

*Grape-sugar tablets.

"Storm" matches, flashlight with a portable generator, sun reflector, first-aid kit

Emergency compass, signal pistol with ammunition, red cloth strips

Three-barreled hunting gun with ammunition (in dust-proof packing), hunting knife

Foreign currency

Emergency radio transmitter

Blood transfusion sets were not included

The search for missing aircraft was undertaken by the desert emergency squadron. It consisted of nine to twelve Storch planes and a number of personnel carriers able to cross desert terrain. The planes patrolled the terrain in question constantly.

British prisoners of war stated that their men had small compasses camouflaged as uniform buttons and handkerchiefs printed with the map of North Africa and southern Europe. This was undoubtedly a very good idea.

25. Dry Docks and Port Installations

The information contained in chapter I, section 1, should be consulted in regard to the practicability and usefulness of the available signal communications material. In the whole of Italian North Africa, there was not a single dry dock, except for small vessels. The port installations of Tripoli, Benghasi, and Tobruk were good, if one considers the conditions prevailing in the colonies. From a technical viewpoint, there were no difficulties in unloading supplies. Difficulties appeared only after the port and unloading installations had been bombed. Besides the above-mentioned three large ports, there were also some smaller unloading points, but they played little or no role in large-scale supply traffic. In Egypt, only Marsa Matruk was used as an unloading port. However, the capacity available was not used in full because of the proximity of the enemy and the bombings from the air.

26. Reinforcement of Sand Surfaces for Landings by Amphibious Craft

The Germans had no experience in reinforcing the sand surfaces on the coast and in the desert for landings by amphibious craft and for landing fields. The Luftwaffe used Italian and British airfields. The maintenance and improvement of roads was in the hands of the Italians. The British used steel matting for their airfields.

Except during the rainy season, the surface of the desert was so hard that reinforcement of roads and airfields was unnecessary. Even the asphalted Via Balbia and the Egyptian coast road had only a foundation of broken limestone without an intermediate layer.

Except for the ports mentioned in section 25, no advanced supply bases were established or used on the open coast. These were unnecessary because not even the normal ports near the enemy lines could be used fully. As a result, no steps were taken to reinforce the beaches for landings by amphibious craft. The whole of the coast consisted of sandy beaches, where the special landing craft of the German Navy, later imitated by the Italian Navy, were able to land. Basically, the difficulties that arose were caused by the sea, which made a landing impossible when the wind blew from the shore.

27. Changes in Ship Loading and Unloading Procedures

The loading and unloading of ships was not influenced by the desert climate. Moreover, in the coastal ports, as has been mentioned, the desert climate did not prevail. In comparison with the interior, the coast was much milder. In the main, native Arabs were used to unload the ships. The limited unloading capacity in comparison with European standards was due to the performance of these people, which was basically lower, and the difficulty in getting them to go back to work after air attack alarms and bombing raids. During the hottest hours, loading and unloading was interrupted when the situation permitted.

28. Materiel Losses and Replacement Estimates for Desert Warfare

It has not been possible to check the accuracy of the German estimates on materiel losses and replacement because a large part of the needs of the forces was covered by captured equipment, the amount of which was not listed.

29. Modifications in Supply-Dump Procedures—Especially for POL (petroleum, oils, and lubricants)

There were no essential changes in the stocking and distribution of fuel. Such changes were unnecessary. Fuel was conserved in German Wehrmacht jerry cans, which proved effective even in desert warfare.

30. Diseases and Insects in the Desert

In 1941, there were a great number of jaundice and dysentery cases. The reason was bowel irritations provoked by the drinking of bad water. Quick treatment with sulfa drugs afforded relief. (The tales that the health of some people was gravely damaged by sand fleas and scorpions are pure fiction.)

The health of the troops requires constant control, which should go hand in hand with thorough instruction. The battalion and other medical officers do not suffice for this purpose. It is, accordingly, advisable to assign a sanitation officer to every company and battery to do this work in close cooperation with a competent medical officer. The sanitation officer should give special attention to a number of protective measures. He will have to see that the soldiers wear warm clothes after sundown and that they do not drink unboiled water. Excretory matter should be buried after each movement of the bowels (the "spade system" is preferable to the latrine system). Whenever the situation permits, the soldiers should observe a siesta. Control of flies should be organized (mosquito nets were used only in the latrines). Vitamin C should be taken regularly. In specified districts, atabrine tablets should be taken to prevent malaria. If necessary, the water supply should be controlled (but only in the case of detached units). There should be a hygienic consultant for every two divisions, and he should be provided with a mobile bacteriological laboratory.

There should be a mobile medical company in each division. The company should have 250 beds available. The German hospitals had the so-called Olympian beds, which proved very useful. In addition, the equipment of a medical company should include one light X-ray instrument and three sets of blood transfusion instruments with plasma. Furthermore, the corps and/or army medical officer will have to have available two surgical hospitals (with 300—400 beds) and two medical companies for every two to three divisions. This is in addition to the medical companies of the divisions.

In respect to insects, there are flies wherever there are people. At first, the troops had no effective means to combat them. It was only during the final phase of the African campaign that the newly developed means (for instance, a preparation called DDT) could be issued. There was no way of preventing infectious diseases, such as dysentery and contagious jaundice, from spreading. The only preventive measure that could be taken was the order for the men to bury all fecal matters immediately. Infringements were severely punished. In contrast to flies, which were everywhere, mosquitoes (anopheles) were present only in the oases and their immediate vicinity. Since these insects fly only at night, use of mosquito nets is an absolute necessity in these areas. Atabrine tablets were used to prevent the malaria transmitted by mosquitoes. Orders specified that one atabrine tablet a day was to be taken. Since the units did not often stay at the oases for long, malaria cases were few.

See chapter III, section 10, for treatment of water.

31. Desert Weather Service

It was the mission of the desert weather service to conduct weather observations and to keep the Luftwaffe posted on the basis of the results of these observations. The information was transmitted by a special radio frequency to a collecting station and retransmitted from there. The mission was carried out by small mobile teams consisting of one or two meteorologists, three or four weather technicians, and several radio operators. A special vehicle contained the weather-service equipment and the radio apparatus (100-watt transmitter and receiver).

The synoptic system was used, with ground weather reports from the numerous ground-observation stations distributed over the coastal area and also from the Italian desert oases. Radio weather-observing stations in Derna sent up two balloons every day. At the front, several pilot-balloon spottings were made, especially early in the morning and in the afternoon (single-station pilot-balloon spotting was done, although it would have been better to use double stations). According to the velocity of the wind at a height of 200—500 meters, it was easy to tell— while the temperature was still low and the wind was weak— whether there would be sandstorms and at what time they would start. Often errors occurred in the wind tests through local causes or because of a rapid change in the atmospheric pressure so that the estimate of sandstorms by means of pilot balloons sent up early in the morning was unreliable. Severe sandstorms will have to be predicted by means of the weather map, with special attention being given to the variations in atmospheric pressure. Small variations in pressure effect considerable changes in the wind at these low latitudes.

In addition, the Luftwaffe transmitted further information over a special radio network. It transmitted individual weather reports, but the main emphasis was on providing fully worked-out weather analyses and surveys and predictions organized by districts. There was a special encoding table for weather reports.

V. GENERAL REMARKS AND EXPERIENCES

32. Special Equipment for Desert Warfare

The proposals for the improvement and modification of equipment—made on the basis of experiences gathered in the desert—were carefully considered by the agencies in the zone of the interior and followed through to the extent that the limited time available permitted. The developments described below are of special importance in desert warfare:

a. Decreasing the sensitivity to dust of vehicular and airplane fuel and of especially sensitive weapons.

b. Experimentation in the production of a dehydrated fuel for vehicles that is also fully effective in saltwater solution. Such a dehydrated fuel greatly alleviates the supply problem and would remove, to a large extent, the danger of air attacks on fuel dumps.

c. Production of radar equipment to identify enemy forces that could not be seen by ground observation.

d. Production of equipment for cooling drinking water or keeping it cool.

e. Procurement of effective smoke-producing equipment that would be capable of sucking up large quantities of desert sand and then releasing it.

f. The mass production of a cross-country truck with a three to three and a half-ton capacity has high priority for supply purposes. The truck will have to possess low-range four-wheel drive and a lock off on the differential. It will need low-pressure balloon tires. The advisability of utilizing air-cooled motors is uncertain. The Steyr Works developed an air-cooled heavy command car, which, however, was not used until shortly before the conclusion of the African campaign so that it could not be adequately tested. Insofar as an evaluation was possible, it seemed to have been in good working order.

Special attention must be accorded the springs. The spiral springs used in German command cars, Types 15 and 17, did not prove practical. The springs broke easily and caused a bottleneck in supply. Leaf springs were better, but they had to be strongly constructed.

All desert vehicles have to be provided with spades, boards, or rope ladders so that they can be dug out when they get stuck and put in action again.

Desert conditions require a sturdy motor vehicle that does not weigh too much, has not got a complicated motor, and has deeply indented tires. Well-equipped replacement dumps and field workshops are a prerequisite. Both must be mobile and capable of producing results quickly.

Camouflage (paint) of all vehicles proved advantageous.

    g. It is advisable to provide all passenger cars and some of the trucks with compasses, which should be so built that they can be easily seen from the driver's seat. The compasses must be compensated by magnets (suitable examples are manufactured by the Askania Works in Berlin-Friedenau).

    h. The field forces must be completely outfitted with tents (with double roofs and dust proofed, where possible). The roof can be flat.

    i. The development of a dry drill similar to the Benoto (see chapter III, section 10, f) that can reach water sources quickly would be very useful. Mobile deep drills should also be at hand, such as are produced in Germany by the Craelius factory.

    j. Since there is a lot more salt water in the desert than fresh water, the field forces can often be supplied, even when no fresh water is available, by having distilling equipment at their disposal. For that reason, it is important to have such equipment. Such equipment should be sparing of fuel and be mounted on cross-country vehicles so that it can be taken along by motorized troops. Its capacity should be great enough to provide each man with four to five liters of water daily.

    k. Ordinary theodolites are adequate for surveying purposes. Still, it is a good idea to provide this equipment with special protection against dust. A military unit should be deployed that can quickly produce reliable maps in cooperation with the air force and can carry out terrain reconnaissance. One can recommend for the last-mentioned mission an organization that is a combination of the English Long Range Desert Group and the German Military Geology Office.

    l. The motors of the Type IV tank (long), which were put into operation in autumn 1941, were very efficient. Depending on the condition of the terrain in which operations had taken place, a general overhauling was necessary after about 10,000—15,000 kilometers. The motors of the Tiger tank, later used in Tunisia, proved too weak and too sensitive. The periscopes of many tanks were often damaged.

The cross-country mobility of the prime movers could have been improved. They were not up to the demands imposed by a thick ground covering of sand.

33. Research and Development Possibilities for Special Desert Equipment

German industrial capacity was capable of satisfying all demands for equipment necessary for desert warfare. It was not prepared for these demands when they came, however. Since technical developments always require considerable time, it was not possible in the comparatively short period during which fighting went on in Africa to technically perfect new equipment. Distribution to the forces did not impose difficulties, as a smooth-working distribution organization was available.

34. Unusual Supply Problems

The basic difficulties in ensuring the smooth-functioning supply of German forces in Africa have already been described in chapter I, section 3.

Overseas theaters of war require many usable unloading ports that are as near to the front as possible. At least partial domination of the sea and air lanes is necessary. Rail connections between unloading ports and the front are essential. In any event, there must be adequate parking space for the supply trains. This space can never be too large, when one considers that all supplies must be sent overland from the harbors to the front. Supply dumps near the front are also necessary and must be kept regularly well stocked.

It is very important that the commander of the army and his chief of staff are daily oriented, in detail, on the supply situation. When waging war in another section of the world, the supply situation must be calculated very accurately, as it exerts an even greater influence on the decisions of the command than elsewhere. On the other hand, the commander should not become overly dependent on the supply situation; nonetheless, he must expect a great effort to be made to get supplies from unloading ports to the front. Only what has already been unloaded can be included in calculations; supplies still on their way cannot be counted.

Tactical considerations will determine the portion of the available ships' tonnage to be used for the transport of men and supplies. This must be directed by an overall authority and is just as true for air transport. The latter can only be of temporary assistance and otherwise must be regarded as a secondary transportation means. The army in the desert, itself, must determine what is to be sent over.

In unloading, care must be taken, because of the danger from air raids, that the ships be emptied as quickly as possible, after which the supplies, especially fuel and ammunition, should be immediately moved from the danger zone into the interior. Strong antiaircraft protection and fighter escorts are indispensable at the unloading ports, as well as barriers against torpedoes and submarines.

Fuel and ammunition must be stored at widely scattered points, where it should be buried and well camouflaged. Burial of materiel is absolutely necessary as protection against the heat. Careful surveillance of camouflage, especially in rear areas, is important.

Rommel's supply columns were not assigned a maximum daily speed or travel time. They were simply given the mission of traveling as rapidly and as long as possible, because a shortage of column space always existed.

There was always a lack of fuel in Africa. After it was unloaded, it was transferred to drums and then issued at the fuel dumps near the front in jerry cans. Underground tank installations are of course desirable, but they did not exist in North Africa at the time.

The average loss of weapons and equipment in the desert was about 25 percent higher than in the European theater of war. Accurate figures, however, are not available (see chapter IV, section 28).

35. Nutrition

Assimilation of food and production of body heat are about 10 to 15 percent lower in hot areas than in Europe. Therefore, under normal conditions, the total nutritional needs are less than in Europe. Heavy physical exertion requires about the same food intake as at home. Too much meat, especially canned meat, was rejected by the men.

A fundamental alteration of nutritional standards soon makes itself felt. Vegetables and fruit are more popular than meat. Too great a meat consumption considerably increases the body's need for water. The high caloric count of fat is in contradiction to the above-mentioned low production of body heat in hot regions. Increased intake of vegetable products, especially marmalade, are necessary to equalize the low fat requirements.

Fresh meat is preferable if refrigeration facilities are available. Smoked meat, especially hard sausage that is not too fatty, finds a ready audience. Eggs should never be eaten raw. Legumes should not be served too often. Moreover, because of the danger of infection with intestinal diseases, greens should be thoroughly washed with a solution of permanganate of potassium. All sorts of dried fruits proved very satisfactory.

The consumption of concentrated alcoholic beverages should be carefully avoided; the best principle is "No alcohol before sunset." The furnishing of lemon and other citric juices is the best means of avoiding Vitamin C deficiency.

As wide a variety as possible should be attempted in the menu, as the troops came to dislike foods that were served constantly.

Refrigeration of foods (meat and fresh vegetables) is achieved by means of so-called cooling units manipulated by aggregates, which can be set up in trucks. The Luftwaffe also had refrigeration planes.

The *Field Cookbook for Improvised Cooking and Baking in the Colonies* was drawn up as a cooking aid.

The crews of armored vehicles and of planes needed extra rations because of the increased pressure exerted upon them.

36. Clothing

Clothing of 100 percent wool is most practical. In general, long trousers that could be tightly closed at the bottom (slacks) were preferred to breeches worn with high-laced boots. The slacks, such as were worn by the Luftwaffe, should be broad and cut on a comfortable bias. This is also true as regards the jackets, the most practical model for which is the Italian Sahariana. A warm cloth coat is indispensable in the cool seasons and nights. Woolen scarves become necessary in winter. Shoes should be constructed of a light leather with a linen inlay and thick, firm soles.

Tents should be painted a striped and speckled green and brown on one side and the color of ocher on the other.

Camouflage nets with knotted strips of many-colored cloth attached to them are the best for sending to the front. Such camouflage nets are practical for all larger motor vehicles.

37. Comparisons with Desert Warfare in Southern Russia

The southern Russian steppes (Kirgiz and Kalmyk steppes) have much the same rainfall as the ring of steppes in North

Africa, 100—200 millimeters a year. However, since the temperature is lower, on the average, evaporation is less. In southern Russia, therefore, the surface of the ground is not covered with a firm crust of surface lime, the vegetation is denser, and watercourses (with water) exist the whole year round—although, with few exceptions, they do not reach the sea but instead filter down through the sand or end in salt lakes. From the military point of view, the differences have the following effect:

    a. The water supply is easier to obtain in the Russian steppes. Although it is not as easy to obtain as in cultivated regions, it nevertheless does not present serious difficulties as long as an organization for reaching it and distributing it exists.

    b. In dry weather, traversability is at least as good as that in the open North African desert. During rainy weather, however, it is worse, since districts with a rocky subsurface are lacking or are to be found only in mountainous areas.

    c. The building of field positions is easier, since the ground is composed of clay or loam almost everywhere.

Like the North African desert, the southern Russian steppes are better suited for attack than for defense. The biggest difficulties are encountered during the severe winter, especially when the enemy troops are accustomed to a harsh climate.

The conditions arising from the dust in Africa are, in general, the same as those prevailing in the pure steppe terrain in southern Russia.

38. Troop Welfare in the Desert

In the African desert, as in no other theater of war, life was hard for officers and men alike. Life in the rear areas, as it is known in Europe, was possible only in the few supply centers in the rear, and then it was only a weak imitation of that elsewhere.

Because of the close contact of the troops with each other, all officers had to give an example of soldierly bearing and good moral living. Any extravagance on the part of an officer is noticed by the men and scrutinized with a magnifying glass. On the other hand, life in the desert offers an excellent opportunity to create a high level of community spirit. This brings about a feeling of solidarity in all ranks and prompts them to do their utmost.

Especially in the desert, where no diversions exist, superiors must be carefully concerned about the welfare of their personnel.

The fact that there are no outside attractions makes it easier for leaders to provide that attention.

An oppressive feeling of immense loneliness overcomes everyone more or less frequently in the desert—a feeling that one is cut off from everything that one holds dear. Commanders must recognize such moods and depressions and offer sincere encouragement so that such pressures will disappear.

Even more important than rations for the well-being of soldiers in the desert is the maintenance of regular communications with the zone of the interior. The word "mail" occupies a place of high priority in desert warfare. All officers and agencies must be concerned, therefore, with assuring a rapid distribution of mail to the front lines.

A good newspaper that carries up-to-the-minute news and gives space to the problems that absorb the soldiers at the front is also indispensable. Movies and theatrical performances at the front can relieve the monotony of the soldiers' lives, even in the desert.

The normal service time in the desert for a German soldier was six months. Regular rotation could, however, often not be provided in Africa between 1941 and 1943, with the result that many soldiers had to serve in the desert for twelve and even eighteen months. If relief is not possible after six months, for one reason or another, the men must be granted three weeks' leave; otherwise, lasting damage to their health can easily result.

Proper leadership, training, and welfare of the troops can lead to extraordinary successes, especially in the seclusion of the desert theater. Field Marshal Rommel has proven this.

Printed in August 2021
by Rotomail Italia S.p.A., Vignate (MI) - Italy